Weather and Water

Developed at
The Lawrence Hall of Science,
University of California, Berkeley
Published and distributed by
Delta Education,
a member of the School Specialty Family

1465667
978-1-62571-172-4
Printing 1 – 6/2014
Quad/Graphics, Versailles, KY

Table of Contents

Readings

Images and Data

References

The Golden Gate Bridge in San Francisco, California, surrounded by fog

Severe Weather

Weather is fairly predictable most of the time. During the summer months in the San Francisco Bay area, you can expect fog in the mornings and late afternoons, with the possibility of sunshine midday. In the southeastern United States, summer days are often hot and humid. In the Midwest and East, winters are usually cold, cloudy, and snowy. These are the normal conditions that people come to expect where they live.

It's the change from normal that catches people's attention, whether they see it on TV or experience it for themselves. **Tornadoes**, **thunderstorms**, windstorms, **hurricanes**, **typhoons**, and **floods** are examples of what is known as **severe weather**. Severe weather is out of the ordinary. It usually causes dangerous conditions that can damage property and threaten lives.

Flood

Floods occur when water overflows the natural or artificial banks of a stream or other body of water. The water moves over normally dry land or accumulates in low-lying areas. Floods often happen with heavy rainfall over short periods of time or when a large amount of snow melts quickly.

Houses destroyed by flood water

Floods may also occur behind ice dams on rivers, during very high tides, or after tsunamis (*soo-nah-mees*, huge waves) caused by earthquakes under water. **Flash floods** are short, rapid, unexpected flows of muddy water rushing down a valley. They are often caused by thunderstorms occurring over mountains, during which large amounts of water flow down a single canyon.

Johnstown, Pennsylvania, 1889. A flood of disastrous proportions hit the city of Johnstown, Pennsylvania in 1889. An earthen dam collapsed after heavy rains. A great wall of water rushed down the Conemaugh River Valley at speeds of up to 64 kilometers (km) an hour. The wall of water, 10 meters (m) high, devastated the town. It washed away most of the northern half of the city, killing 2,209 people and destroying 1,600 houses.

The flood in Johnstown, Pennsylvania, destroyed many houses.

Drought

Droughts are less-than-normal precipitation over a long period of time. They usually cause water shortages, including low flow or no flow in streams. Soil moisture and **groundwater** levels decrease. Droughts are especially disastrous to agriculture.

Drought

The Dust Bowl, United States, 1930s. An area of the United States that includes parts of Colorado, Kansas, New Mexico, and the panhandles of Texas and Oklahoma was named the Dust Bowl in the early 1930s. A severe drought occurred after years of poor

Hail can be very large, causing damage when it falls.

land management. The native grasses had been removed, exposing the topsoil. Strong spring winds blew away the topsoil, causing “black blizzards.” Thousands of families left the area at the height of the Great Depression (1929–1939) because of the poor living conditions brought on by the drought.

A dust storm in the Dust Bowl.

Pakistan, 2000. Large parts of Pakistan were hit by drought in 2000. No rain fell in the Balochistan area for more than 3 years. Many people were forced to leave some of the more remote villages. They migrated to cities in search of food and water. This required a major support effort by the Pakistan government and several international relief agencies.

Hail

Hail is precipitation in the form of balls of ice. The diameter of hailstones ranges from 0.5 to 10 centimeters (cm). Hail usually forms during thunderstorms when strong updrafts (vertical winds) move through cumulonimbus clouds in which temperatures are near or below freezing.

Northwestern Missouri, 1898. A huge hailstorm struck Nodaway County in northwest Missouri on September 5, 1898. The hail remained on the ground for 52 days and left the fields unworkable for 2 weeks. On October 27, there was still enough hail left in ravines for the local residents to make ice cream.

Hail creates dangerous road conditions.

Selden, Kansas, 1959. On June 3, 1959, a severe hailstorm struck the town of Selden in northwestern Kansas. Hail fell for 85 minutes. The storm covered an area 10 km by 14.4 km with hailstones to a depth of 46 cm. The storm caused $500,000 worth of damage. That was a lot of money in 1959.

Hurricane

Hurricanes (also called cyclones) are moving wind systems that rotate around an eye, or center of low **atmospheric pressure**. Hurricanes form over warm tropical seas. Wind speeds are more than 119 km per hour in a hurricane. The term *hurricane* is used for Northern Hemisphere cyclones east of the International Date Line to the Greenwich Meridian. The term *typhoon* is used for Pacific cyclones north of the equator west of the International Date Line. Hurricanes and typhoons produce dangerous winds, heavy rains, and flooding. They can cause great property damage and loss of life in coastal areas.

Galveston, Texas, 1900. The hurricane that struck Galveston, Texas, on September 8, 1900, is considered to be one of the worst natural disasters in US history. More than 6,000 men, women, and children lost their lives during this great storm. It is estimated that the winds reached 250 to 333 km per hour. The tidal surge was probably 4.6 to 6.2 m. This towered over the highest elevation on Galveston Island, which was 2.7 m. More than 3,600 houses were destroyed, with whole blocks of houses totally wiped out, leaving only a few bricks behind.

Galveston Island after the 1900 hurricane

Hurricanes are defined by strong winds.

New Orleans, Louisiana, 2005. Hurricane Katrina was the costliest hurricane in the history of the United States, and was the sixth strongest Atlantic hurricane on record. Damage stretched along the Gulf Coast from Texas to central Florida. The levee system failed in New Orleans, leaving the city vulnerable to the incredible storm surge and flooding. More than 1,800 lives were lost, and more than one million people were displaced from their homes.

Lightning and Thunder

Lightning is a visible electric discharge produced by thunderstorms. Not everyone agrees *why* lightning happens, but *what* happens is pretty well understood. Lightning travels from a cloud to the ground. The electric charge moves downward in steps of approximately 46 m called **step leaders**. This flow continues until the charge reaches something on the ground that is a good conductor. The circuit is completed.

The whole event usually takes less than half a second, and a lightning bolt can reach 200 million volts.

Since 1989, US meteorologists have used a network of antennas to detect lightning. An average of 20 million cloud-to-ground strikes happen every year over the continental United States.

Thunder is the explosive sound that usually accompanies lightning. The sound is caused by the rapidly expanding gases in the **atmosphere** along the path of the lightning. How loud and what type of sound you hear depends on atmospheric conditions and how far away you are from the flash.

Lightning

Durham, North Carolina, 1995. Lightning during an early-evening thunderstorm killed a golfer in Durham, North Carolina, on July 7, 1995. The man took cover from the storm in a wooden shed at a golf course. Witnesses say the lightning first struck a tree and then bounced to the shed. Other golfers sustained scrapes and burns from the accident. Later they said that the lightning was the brightest they had ever seen and that the thunder came immediately.

Villa Gesell, Argentina, 2014. Three people were killed and 22 others were injured when lightning struck a beach during a sudden thunderstorm. Some people tried to take shelter in a tent, but the structure was not protective enough and they sustained injuries.

The fast-moving column of air in a tornado can cause a lot of damage.

Tornado

Tornadoes are rapidly rotating columns of **air** that extend from a thunderstorm to the ground. Wind speeds in a tornado can reach 417 km per hour or more. The path of a tornado can be more than 1 km wide and 83 km long.

Xenia, Ohio, 1974. On April 3 and 4, 1974, a super tornado outbreak struck the states of Alabama, Georgia, Illinois, Indiana, Kentucky, Michigan, Mississippi, North Carolina, Ohio, South Carolina, Tennessee, Virginia, and West Virginia. Especially devastating tornadoes struck Ohio during the afternoon and early evening of April 3. The town of Xenia was hit the hardest. Thirty people died, more than 1,100 were injured, and more than 1,000 houses were destroyed. The path of damage ranged from 0.4 to 0.83 km wide.

Oklahoma City, Oklahoma, 1999. Eight thunderstorms produced at least 59 different tornadoes in central Oklahoma alone on May 3, 1999. Many of these tornadoes were very violent and long-lasting. They made direct hits on several populated areas, including Oklahoma City. At least 40 people died in Oklahoma because of the twisters, and 675 were injured. The total damage was about $1.2 billion. The tornadoes also caused extensive damage to the Wichita, Kansas, metro area.

Enterprise, Alabama, 2007. A tornado outbreak spread across the southern United States for 3 days in March 2007, with a total of 56 confirmed tornadoes. Enterprise, Alabama, was hit the hardest. Tragedy struck when the tornado went directly through a high school, causing eight fatalities, collapsing part of the school, flipping cars in the parking lot, and tearing trees from the ground. The tornado was estimated to have been 457 m in diameter and to have traveled for 16 km.

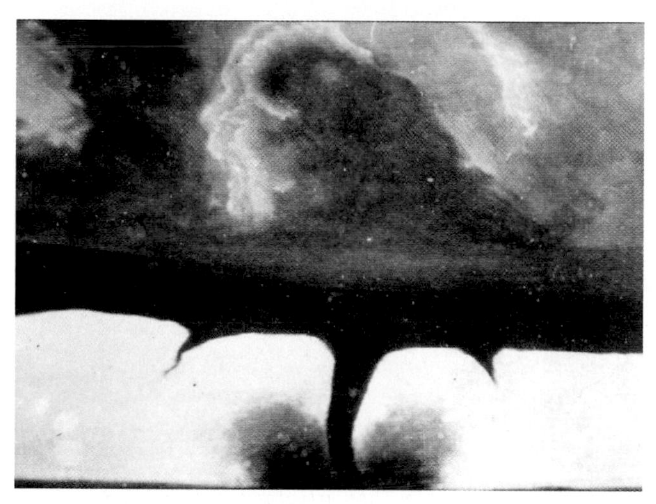

This is the oldest known photo of a tornado. It was taken 35 km southwest of Howard, South Dakota, on August 28, 1884.

Thunderstorm

Thunderstorms produce rapidly rising air currents, usually resulting in heavy rain or hail along with thunder and lightning. A thunderstorm is classified as severe when it produces one or more of the following:

- Hail 1.9 cm or greater in diameter
- Winds gusting in excess of 96 km per hour
- A tornado

Thunderstorms often result in heavy rain.

West Central Texas, 1996. On October 22, 1996, surface temperatures dropped from 10 degrees Celsius (°C) to around 0°C in an area of west central Texas. Strong rising winds developed. The storm increased in intensity very quickly. Dime- to egg-sized hail fell, as well as large amounts of sleet and light snow.

Fort Worth, Texas, 1995. An isolated severe storm became the costliest thunderstorm in US history when it devastated the area in and around Fort Worth, Texas, on May 5, 1995. More than 100 people were injured, mostly by softball-sized hail that pelted people attending an outdoor festival. Winds reached 100 to 117 km per hour, dropping hail the size of grapefruit in some areas. The large hail and high winds damaged hundreds of houses, businesses, and vehicles. Damage totals reached more than $2 billion.

Blizzard

Blizzards are severe storms with low temperatures, strong winds, and large amounts of snow. Blizzards have winds of more than 51 km per hour and enough snow to limit visibility to 150 m or less. A severe blizzard has winds of more than 72 km per hour, near zero visibility, and temperatures of –12°C or lower.

East Coast, United States, 1899. A cold wave hit the East Coast from February 1–14, 1899, causing a huge blizzard and bitter-cold temperatures from the Rockies to the Atlantic Ocean. Snow fell from Louisiana to Georgia and extended northeast into New England. Up to 61 cm of snow fell across much of the Northeast. Winds gusted to 58 km per hour. Florida experienced its first recorded blizzard. Temperatures fell below freezing. Jacksonville, Florida, measured its greatest snowfall ever at just over 4 cm. Tampa, Florida, received its first measurable snow on record.

In blizzards, snow can fall rapidly and winds can blow it into huge drifts that are taller than humans.

Blizzards decrease visibility, making travel difficult and dangerous.

Snow and ice can bring down power lines.

Chicago, Illinois, 1967. Chicago, Illinois, set records on January 26 and 27 with the worst blizzard ever recorded there. More than 52 cm of snow fell in 29 hours and 8 minutes. Trains froze and could not move. Buses were stranded, and O'Hare Airport was closed for 3 days.

Washington, DC, 1979. Known as the Presidents' Day Storm, this blizzard covered much of the Northeast, closing down Washington, DC. Ice formed on the Potomac River. Some mid-Atlantic areas received more than 44 cm of snow. Observers in Baltimore, Maryland, reported snow falling at a rate of 11 cm per hour.

Super Storm, 1993. Twenty-six states were affected by a low-pressure system during March 12–15, 1993. The system brought heavy snow and strong winds. Birmingham, Alabama, received 28.6 cm of snow, while Chattanooga, Tennessee, measured 44 cm by the time the storm passed. Winds gusted up to 83 km per hour as temperatures fell rapidly. Parts of Long Island, New York, recorded wind gusts of 155 km per hour. Over 50 tornadoes hit Florida. The storm claimed more than 200 lives.

A dust devil

Windstorm

Sometimes strong winds that are not directly connected with thunderstorms, tornadoes, or hurricanes occur. **Straight-line winds** are strong winds that have no **rotation**. The winds can travel at speeds of more than 167 km per hour. If no one is around to observe what happens, it can be difficult to tell if damage came from a straight-line wind or a tornado.

Microbursts are small, very intense downward winds. They affect areas less than 4 km wide. Microbursts can produce winds of more than 280 km per hour. They typically last less than 10 minutes. They are often associated with thunderstorms. Microbursts often cause significant ground damage and are a threat to aviation.

Dust devils are small rotating winds not associated with a thunderstorm. They become visible when they collect dust or debris. Dust devils form when air is heated by a hot surface during fair, hot weather. You are most likely to see a dust devil in arid or semiarid regions such as Texas or New Mexico.

Dust storms are rare conditions in which strong winds carry dust over a large area. They occur when there are drought conditions. In desert areas, sandstorms occur.

A dust storm advances across the land.

Wildfires are driven by winds. Windstorms can spread fire across large areas, making the fire harder to extinguish. Many regions in the world, such as Australia and the southwestern United States, are susceptible to wildfires, so windstorms carry additional risks in these areas.

China, 1998. Parts of China and Mongolia experienced a major dust storm in April 1998. Twelve people were reported missing after the storm, which brought very strong winds. The storm blanketed ten cities and districts. Power and water supplies were cut. A trace of yellow dust carried by the winds was deposited on nearby deserts. Some of the dust was blown across the Pacific Ocean to the United States.

Mars, 2012. Wind and dust can be severe on Mars. A large Martian dust storm can cover most of the **planet**. NASA's Mars Reconnaissance Orbiter photographed a dust devil blowing across Mars's northern plains in February 2012. The dust devil in the photograph below has a diameter of 30 m.

What types of severe weather occur where you live? Record a list in your notebook.

A dust devil on Mars

Naming Hurricanes

October 29, 2012—Sandy moves toward the eastern seaboard of the United States with winds of 100–200 kilometers (km) per hour and a storm surge over 9 meters (m) high. Its effect is forecasted to reach from Florida to Maine, affecting the entire eastern US coastline.

Who is Sandy? And who are Katrina, Camille, Hugo, and Andrew? They're hurricanes! Throughout human history, hurricanes have caused destruction. Some of the most powerful hurricanes to make landfall in the United States were those named above. How did these storms get names, and why?

Because they have reputations as troublemakers, hurricanes are under the watchful eye of meteorologists from the time they first take form as tropical storms. In order to refer to them over a period of days and weeks, they are given names. It's much easier to talk about what Gertie did today than "that hurricane 18°N, 55°W." It also helps reduce confusion when more than one hurricane is brewing in the same area.

The practice of naming hurricanes has undergone a number of changes over the years. In the West Indies (perhaps in hopes of protection), hurricanes were first named after the saint whose day fell closest to the arrival of the hurricane. Early in the 20th century, an Australian forecaster started naming hurricanes for politicians. People thought this was funny because news reports could include headlines like "Dundee causing great distress" or "Gibson wandering aimlessly, but could be dangerous."

Hurricanes first take form as tropical storms.

For a few years, the joint army/navy phonetic alphabet (Able, Baker, Charlie . . .) was used to name hurricanes. The first hurricane of the **season** was Able, Baker was second, and so on. During World War II, US Air Force and Navy meteorologists named the Pacific storms for their wives or girlfriends. Starting in 1953, meteorologists in the United States used whatever names they liked as long as they were women's names, and the first hurricane of the season had a name starting with the letter *A*.

In 1977, the World Meteorological Organization, a United Nations group, decided on the official naming system used today. They came up with 6 years' worth of names for storms in the Atlantic Ocean. After 6 years, they start over at the beginning. The six lists are alphabetical, and the letters *Q*, *U*, *X*, *Y*, and *Z* are not used. The first forecaster who came up with 21 names used a baby-naming book. He added names of his relatives to the list, too. It hasn't happened yet, but if all 21 names on a year's list are used up, additional storms would be assigned Greek letters, like Alpha and Beta.

The first versions of the lists contained only female names. In 1978, the lists were changed to include both male and female names. They also include French and Spanish names for Atlantic storms.

Outside the Atlantic, 14 countries of the western North Pacific and South China Sea have approved lists of names for tropical storms originating in the South Pacific. The names derive from many languages, and include Nakri (a type of flower in Cambodia), Haishen (a Chinese sea god), Pabuk (a big freshwater fish in Laos), Parma (a dish of ham, liver, and mushrooms popular in Macao), Ewiniar (a Chuuk storm god in Micronesia), Cimaron (a wild ox in the Philippines), Durian (a Thai fruit), and Con Son (a mountain in northern Vietnam).

Damage caused by Hurricane Katrina

The names of the most severe storms are taken off the list for at least 10 years. At this time, the retired names include Katrina, Sandy, Andrew, and Hugo. The country that suffered the most destruction from the hurricane gets to add a new name.

When do storms get a name? Tropical storms are named once they start rotating and reach a wind speed above 65 km per hour. A tropical storm becomes a hurricane when its winds reach 119 km per hour, and it keeps the same name.

Storm Names for 2014–2019 (Atlantic Storms)

2014	2015	2016	2017	2018	2019
Arthur	Ana	Alex	Arlene	Alberto	Andrea
Bertha	Bill	Bonnie	Bret	Beryl	Barry
Cristobal	Claudette	Colin	Cindy	Chris	Chantal
Dolly	Danny	Danielle	Don	Debby	Dorian
Edouard	Erika	Earl	Emily	Ernesto	Erin
Fay	Fred	Fiona	Franklin	Florence	Fernand
Gonzalo	Grace	Gaston	Gert	Gordon	Gabrielle
Hanna	Henri	Hermine	Harvey	Helene	Humberto
Isaias	Ida	Ian	Irma	Isaac	Ingrid
Josephine	Joaquin	Julia	Jose	Joyce	Jerry
Kyle	Kate	Karl	Katia	Kirk	Karen
Laura	Larry	Lisa	Lee	Leslie	Lorenzo
Marco	Mindy	Matthew	Maria	Michael	Melissa
Nana	Nicholas	Nicole	Nate	Nadine	Nestor
Omar	Odette	Otto	Ophelia	Oscar	Olga
Paulette	Peter	Paula	Philippe	Patty	Pablo
Rene	Rose	Richard	Rina	Rafael	Rebekah
Sally	Sam	Shary	Sean	Sara	Sebastien
Teddy	Teresa	Tobias	Tammy	Tony	Tanya
Vicky	Victor	Virginie	Vince	Valerie	Van
Wilfred	Wanda	Walter	Whitney	William	Wendy

Hurricane Sandy caused widespread devastation on the East Coast of the United States, particularly in New York and New Jersey.

Mr. Tornado

"Grabbing a pencil and paper, I rushed to the rooftop where I began recording the direction of cloud-to-ground lightning and the time between the flash and subsequent thunder."

That's how Professor Tetsuya Theodore Fujita (1920–1998) described the beginning of his lifelong fascination with severe weather. He was 27 years old when he made those first observations. After completing his degree in meteorology at the University of Tokyo, Fujita joined the faculty at the University of Chicago. Some of the most powerful storms in the world occur in the stretch of land that runs north and east from Texas through Oklahoma and Kansas, and into the Midwest. This corridor, known as Tornado Alley, was where Fujita focused much of his research over the next several decades.

Shortly after his arrival in Chicago, Fujita began analyzing individual thunderstorms in detail. He observed and recorded temperature, **air pressure**, and wind data and related them to the development of the huge, dark clouds that form during thunderstorms. His interest in thunderstorms led directly to tornadoes. It was in the area of tornado science that Fujita established his reputation as a pioneer researcher. Among his peers he was nicknamed Mr. Tornado, in recognition of his great contributions.

Fujita introduced the concept of tornado families, a group of individual tornadoes, each with a unique path, spawned by the same massive thunderstorm. Prior to this, people thought that long damage paths were made by a single tornado. Fujita discovered that, as a thunderstorm advanced, two, three, or more funnels might form, touch down, break up, and re-form later to create more destruction on the ground. On occasion, two or more funnels might extend from the storm front at the same time. He introduced new concepts of thunderstorm architecture and developed terms like *wall cloud* and *tail cloud*.

Tornado damage in Murfreesboro, Tennessee

In the late 1960s, Fujita's analysis of the Palm Sunday tornadoes in the Midwest in 1965 changed the course of how we view a tornado outbreak. For the first time, he mapped the entire outbreak in terms of tornado families. While multiple-vortex tornadoes are well known today, he was the first to identify their existence, based on damage patterns.

In the 1970s, Fujita again revolutionized tornado science by giving us a system that linked damage and wind speed. Previously, all tornadoes were counted as equals. Fujita quantified their force on a five-level scale, called the Fujita scale, with force 5 (F5) being the most powerful and potentially destructive storms. For the super outbreak of 1974 in the central and eastern United States, Fujita was able to develop scale-intensity contour maps for the entire path of many of the 148 tornadoes that raged that year. After 40 years, we still use his ideas and terminology.

In the late 1970s, Fujita turned his attention to weather-related aircraft disasters. He identified two phenomena that had not been described before, the **downburst** and the microburst. Before this, meteorologists had been confused by a bewildering array of gusty winds in and around thunderstorms. By the 1980s, downburst and microburst were understood as separate kinds of winds. They were known to be caused by events such as dry air moving into a thunderstorm.

In Fujita's later years, he applied his knowledge to hurricanes and typhoons, which are the kind of severe weather that he had originally focused on in Japan, where he was born.

Do you have an area of interest that might become a part of your future career? Record your ideas in your notebook.

Traditional Weather Tools

Many modern weather tools use digital sensors, like the weather station you have in class. Here are some of the more traditional weather tools you may have seen.

Thermometer

A **thermometer** measures temperature. Temperature is measured in degrees. This thermometer uses the Celsius scale (°C). You would read the temperature as 25°C in this photo.

Some thermometers use the Fahrenheit scale (°F), while many feature both scales.

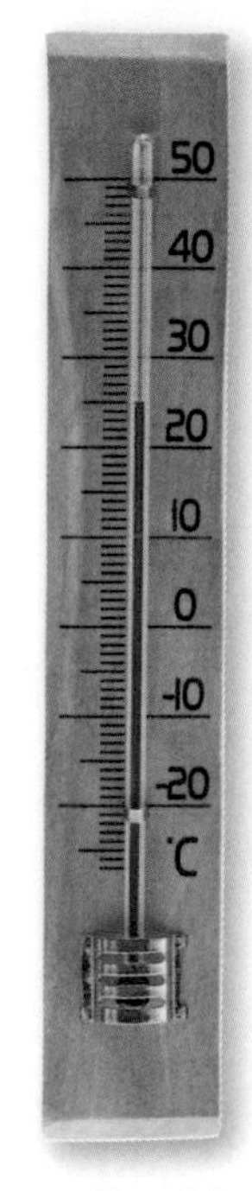

A liquid thermometer

Compass

Use a **compass** to determine the direction from which the wind is coming.

- Stand facing the wind.
- Hold the compass in front of you.
- Rotate the compass body until the *N* lines up with the needle.
- A line running from the center of the compass needle straight into the wind tells you the wind direction.

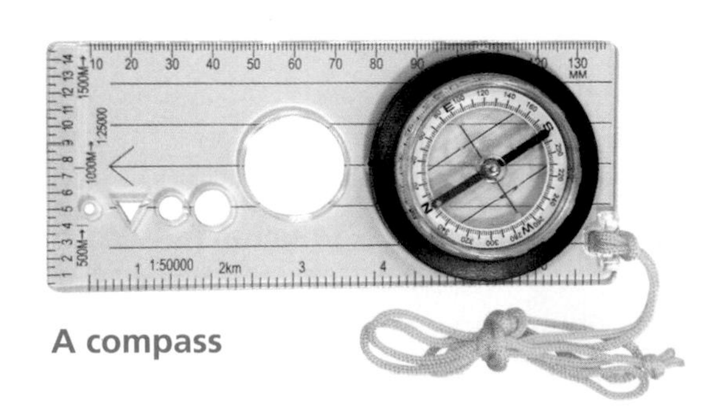

A compass

Wind Meter

Use a **wind meter** to measure wind speed.

- Stand facing the wind.
- Hold the meter in front of you straight up and down, scale toward you. Don't cover the holes at the top.
- The height of the ball indicates the wind speed.

If the wind is strong, cover the top hole with your finger to read the high scale on the right.

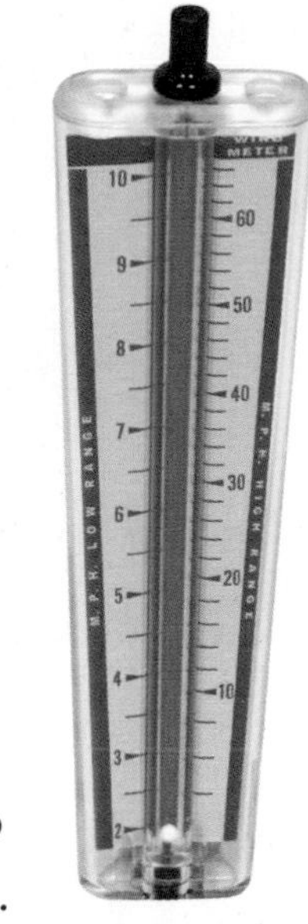

A wind meter

Hygrometer

A **hygrometer** measures relative humidity. Relative humidity is a measure of the amount of **water vapor** in the air. It is a percentage. On the hygrometer in this photo, relative humidity is 58 percent.

A hygrometer

Barometer

A **barometer** measures air pressure. Air pressure on this barometer is measured in millibars (mb). The barometer in this photo indicates a pressure of 1,021 mb.

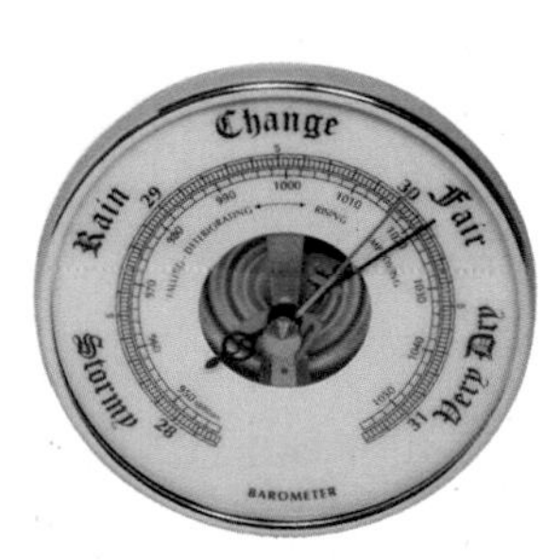

A barometer

What's in the Air?

Air? Can't see it. Can't taste it. Can't smell it. If you pay attention, you might feel it as a gentle breeze brushing across your skin. Then again, a strong gust of wind can nearly knock you over. Because we are typically so insensitive to air, it is difficult to understand what it is. Is it one thing, or a mixture of things? And where is it? Is it everywhere, or just in some places?

As we go about our everyday business, we usually travel with our feet on the solid Earth and our heads in the atmosphere. The atmosphere completely surrounds us, pressing firmly on every square centimeter of our top, front, back, and sides. Even if we attempt to get out of the atmosphere by locking ourselves inside a car or hiding in a basement, the atmosphere is there, filling every space we enter.

An atmosphere is the layer of gases that surrounds a planet or **star**. All planets and stars have an atmosphere around them. The Sun's atmosphere is hydrogen. Mars has a thin atmosphere of **carbon dioxide (CO_2)** with a bit of **nitrogen (N_2)** and a trace of water vapor. Mercury has almost no atmosphere at all. Venus, almost the same size as Earth, has an atmosphere composed of carbon dioxide and nitrogen, which would be toxic to a visiting human. Each planet is surrounded by its own mixture of gases.

Earth's atmosphere is composed of a mixture of gases we call **air**. Air is mostly nitrogen (78 percent) and **oxygen (O_2)** (21 percent), with some argon (0.93 percent), carbon dioxide (0.03 percent), **ozone (O_3)**, water vapor, and other gases (less than 0.04 percent together).

Sketch a pie chart in your notebook to represent the percentage of gases in the atmosphere.

Nitrogen is the most abundant gas in our atmosphere. It is a stable gas, which means it doesn't react easily with other substances. When we breathe air, the nitrogen goes into our lungs and then back out unchanged. We don't need nitrogen gas to survive, but it doesn't harm us either.

Oxygen is the second most abundant gas. It takes up about 21 percent of the air's volume, and because the oxygen atom is larger than the nitrogen atom, it accounts for 23 percent of air's **mass**. Oxygen is a colorless, odorless, and tasteless gas. It is the most plentiful element in the rocks of Earth's crust. Oxygen combines with hydrogen to form water.

Gases of the Atmosphere	
Gas	**Percentage by volume**
Nitrogen	78.08
Oxygen	20.95
Argon	0.93
Water vapor*	0.25
Carbon dioxide*	0.039
Ozone*	0.01
Neon	0.002
Helium	0.0005
Krypton	0.0001
Hydrogen	0.00005
Xenon	0.000009

*Variable gas

Oxygen and nitrogen are called **permanent gases**. The amount of oxygen and nitrogen in the atmosphere stays constant. Most of the other gases in this chart are also permanent gases, but are found in much smaller quantities.

Air also contains **variable gases**. The amount of a variable gas changes in response to activities in the environment. Each of the variable gases listed in the table on this page is also considered to be a **greenhouse gas**, which means it traps heat within the atmosphere. The way these gases trap heat will be studied later in this course. Remember that these gases are variable, so if the amount of greenhouse gases increases, the atmosphere may trap more heat. But greenhouse gases are important for life on Earth! If we didn't have greenhouse gases, the Earth would be too hot during the day and too cold during the night to support life.

Water vapor (H_2O) is the most abundant variable gas. It makes up about 0.25 percent of the atmosphere's mass. The amount of water vapor in the atmosphere changes constantly. Water cycles between Earth's surface and the atmosphere through evaporation, condensation, and precipitation. You can get a feeling for the changes in atmospheric water vapor by observing clouds and noting the stickiness you feel on humid days.

Carbon dioxide is another important variable gas. It makes up only about 0.04 percent of the atmosphere. You can't see or feel changes in the amount of carbon dioxide in the atmosphere.

Carbon dioxide plays an important role in the lives of plants and algae. These organisms remove carbon dioxide from the air during **photosynthesis**. Plants and algae convert light energy into chemical energy by making sugar (food) out of carbon dioxide and water. This process releases oxygen into the atmosphere. When living organisms use the energy in food to stay alive, they remove oxygen from the air and return carbon dioxide to the air.

There are other gases that you may have heard about. Ozone (O_3) is a variable gas. It is a form of oxygen that accumulates in the **stratosphere**. Ozone is absolutely essential to life on Earth because it **absorbs** deadly ultraviolet radiation from the Sun. But ozone in high concentration can cause lung damage. In the lower atmosphere, ozone is an air pollutant.

Methane (CH_4) is a variable gas that is increasing in concentration in the atmosphere. Scientists are trying to figure out why this is happening. They suspect several things. Cattle produce methane in their digestive processes. Methane also comes from coal mines, oil wells, and gas pipelines, and is a by-product of rice cultivation. Methane, like other greenhouse gases, absorbs heat coming up from Earth's surface.

These gases and a few other trace gases are all mixed together, so that any sample of air is a mixture of all of them. If you rise higher in the atmosphere, there are fewer particles, but the ratio of each gas to the other is the same. The mixing is caused by the constant movement of the air in the part of the atmosphere near Earth's surface. Above about 90 kilometers (km), there is much less mixing. Very light gases (hydrogen and helium, in particular) are more abundant above that level.

Think Questions

1. What is the difference between permanent gases and variable gases in the atmosphere?
2. During the daylight hours, plants and algae take in carbon dioxide and release oxygen. If humans remove forests, what might happen to the balance between these gases?

Methane is generated through different processes, including the digestive processes of cattle.

A Thin Blue Veil

Space-shuttle astronauts took this photo while orbiting Earth. You can see a side view of Earth's atmosphere. The black bumps pushing into the troposphere are tall cumulus clouds.

The crew of Apollo 17 took this photograph of Earth in December 1972, while on their way to the Moon. The small box at the top of this photo shows an area equal to the atmosphere photo above taken by space-shuttle astronauts.

It is cold in deep space. The temperature is in the neighborhood of –270 degrees Celsius (°C). That's nearly 200°C colder than it has ever been on Earth. Near stars, like the Sun, it's outlandishly hot, reaching thousands of degrees. There are, however, a few places here and there in the universe where the temperature is between the extremes. Earth is one of those places. The average temperature on Earth is not too hot and not too cold, just right for supporting life.

On a typical day, the temperature range on Earth is only about 100°C, from maybe 45°C in the hottest place to –55°C at one of the poles. The extremes are 57°C in Furnace Creek Ranch, Death Valley National Park, California, recorded on October 7, 1919, and –89°C in Vostok, Antarctica, on July 21, 1983. That's a range of temperature on Earth of 146°C.

It's not only because we are at the right distance from the Sun that Earth has tolerable temperatures. Earth's atmosphere keeps the temperature within a narrow range that is suitable for life.

From space, Earth's atmosphere looks like a thin blue veil. Some people call it an ocean of air. The depth of this "ocean" is about 600 kilometers (km). The atmosphere is most dense right at the bottom where it rests on Earth's surface. The air gets thinner and thinner (less dense) as you move away from Earth's surface, until it disappears.

Imagine a column of air that starts on Earth's surface and extends up 600 km to the top of the atmosphere. Scientists have discovered several distinct layers in this column of air. Each layer has a different temperature. Here's how it stacks up.

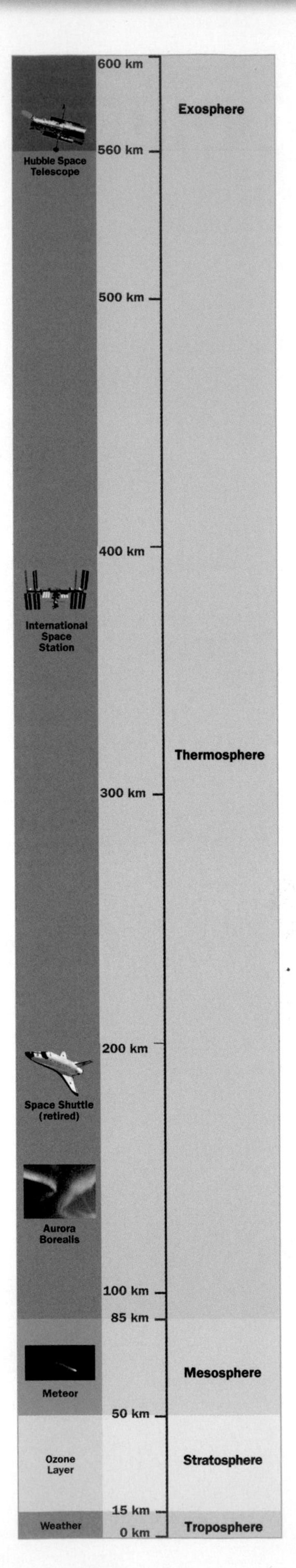

Earth's atmosphere

The layer we live in is the **troposphere**. It starts at Earth's surface and extends upward for 10–24 km. Its thickness depends on the season and where you are on Earth. Over the warm equator, the troposphere is a little thicker than it is over the polar regions, where the air is colder. It also thickens during the summer and thins during the winter. A good average thickness for the troposphere is 15 km.

This ground-floor layer contains most of the organisms, dust, water vapor, and clouds found in the entire atmosphere. For that matter, it contains most of the air as well. And most important, weather happens in the troposphere. The troposphere is where the action is. This is where differences in air temperature, humidity (moisture), air pressure, and wind occur.

Weather occurs in the troposphere.

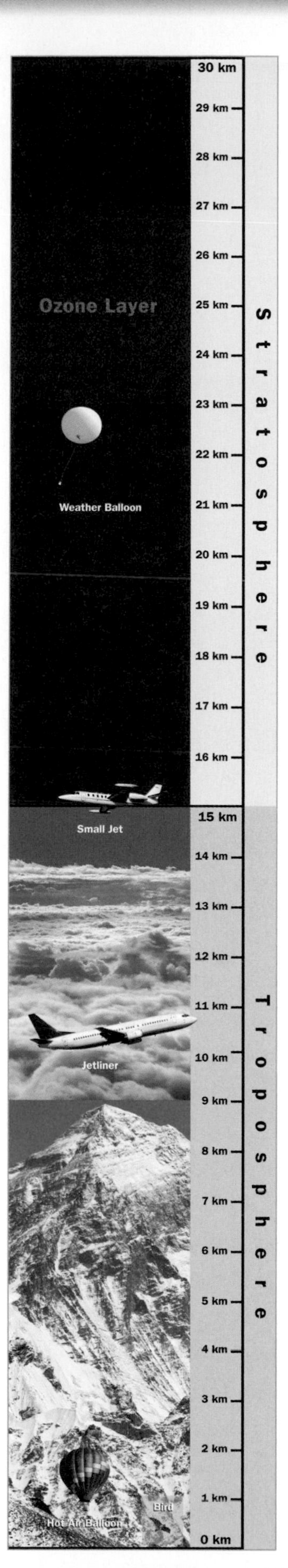

The first 30 km of the atmosphere

Weather balloons are launched to monitor temperature, humidity, air pressure, and wind.

These properties of air temperature, humidity, air pressure, and wind are called **weather factors**. Meteorologists launch weather balloons twice a day to monitor weather factors. The balloons float up through the atmosphere to about 23 km. Weather factors will be investigated in detail as we continue to study weather.

The troposphere is the thinnest layer, only about 2 percent of the depth of the atmosphere. It is the most dense layer, however, containing four-fifths (80 percent) of the total mass of the atmosphere.

Earth's surface (land and water) absorbs heat from the Sun and warms the air above it. Because air in the troposphere is heated mostly by Earth's surface, the air is warmest close to the ground. The air temperature drops as you go higher. At its upper limit, the temperature of the troposphere is about –60°C. The average temperature of the troposphere is about 25°C.

Mount Everest, located in Nepal and Tibet, is the highest landform on Earth, rising 8.8 km into the troposphere. The air temperature at the top of the mountain is well below freezing most of the time. There is also less air to breathe at the top of Mount Everest. Climbers usually bring oxygen along to help them survive the thin air.

The stratosphere is the layer above the troposphere. It is 15–50 km above Earth's surface and contains almost no moisture or dust. It does, however, contain a layer of ozone (O_3), a form of oxygen, that absorbs high-energy ultraviolet (UV) radiation from the Sun. The temperature stays cold until you reach the upper reaches of the stratosphere, where energy absorption by ozone warms the air to about 0°C.

Mount Everest

The **jet stream**, a fast-flowing river of wind, travels generally west to east in the region between the lower stratosphere and the upper troposphere. Many military and commercial jet aircraft take advantage of the jet stream when flying from west to east.

The **mesosphere** is above the stratosphere, 50–85 km above Earth's surface. The temperature plunges again, reaching its coldest temperature of around –90°C in the upper mesosphere. This is the layer in which meteors burn up while entering Earth's atmosphere, producing what we call shooting stars.

Beyond the mesosphere, 85–560 km above Earth, is the **thermosphere**. The thermosphere is the least-understood layer of the atmosphere and the most difficult to measure. The air is extremely thin. The thermosphere is the region of the atmosphere that is first heated by the Sun. A small amount of energy coming from the Sun can result in a large temperature change. When the Sun is extra active with sunspots or solar flares, the temperature of the thermosphere can surge up to 1,500°C or higher!

Within the thermosphere are regions noted for their chemistry properties. These regions contain a large number of electrically charged ions. Ions form when intense radiation from the Sun hits particles in the atmosphere. These ionized particles are responsible for the aurora borealis, or northern lights, and the aurora australis, or southern lights.

The identification of these four layers (troposphere, stratosphere, mesosphere, and thermosphere) is based on temperature. There are no sharp boundaries or abrupt changes in gas composition between them. As average temperatures change with the seasons, the boundaries between layers may move up or down a little.

The Northern lights (aurora borealis)

Beyond the thermosphere, Earth's atmosphere makes a transition into space. This area is the **exosphere**, where particles from the atmosphere escape into space. It extends from the top of the thermosphere up to 10,000 km. In this region, the temperature plunges to the extreme –270°C of outer space, and the concentration of atmospheric gases fades to nothing.

That 600 km column of air that makes up the atmosphere pushes down on the surface of Earth with a lot of force. We call this force air pressure, or atmospheric pressure. We are not aware of it because we are adapted to live under all that pressure, but there is a mass of about 1 kilogram (kg) pushing down on every square centimeter of surface on Earth.

Here's another way to look at it. If all the mass of the air were replaced with solid gold, the entire planet would be covered by a layer of gold a little more than half a meter deep. That's a lot of gold, but the atmosphere is much more valuable.

Think Questions

1. How is Earth's atmosphere like the ocean? How is it unlike the ocean?
2. Why do you think airplanes don't fly high in the stratosphere?

The International Space Station orbits in the thermosphere.

What Is Air Pressure?

The atmosphere is composed of air. Air has mass. In fact, a column of air 1 square centimeter (cm^2) in area extending up to the top of the atmosphere has a mass of 1.2 kilograms (kg). If the top of your head has a surface area of 150 cm^2, that means every time you go out and stand under the open sky, you have the pressure of 180 kg of air pushing down on your head! That's like wearing a hat with a refrigerator on it. Is it safe to go outside?

Don't worry, it's safe. The force applied by the air above you is called atmospheric pressure. Life on Earth has evolved in this high-pressure environment, so we are able to handle the pressure just fine. In fact, most of the time we are totally unaware of the pressure.

We do feel the air pressure when it changes quickly. Have you ever traveled to the mountains and noticed a popping in your ears as you go up or down the mountain? Sometimes this happens in an airplane when it changes altitude rapidly or in a car going up and down a hilly road. If you have had that experience, keep it in mind as you think more about atmospheric pressure.

Sometimes you can see evidence of change in pressure even if you can't feel it. In an airplane, you might notice that packets of peanuts or chips are puffed up like balloons.

This bottle of air appeared uncrushed when it was sealed at high altitude. Then it was brought to sea level. Its flattened look is evidence of changed air pressure.

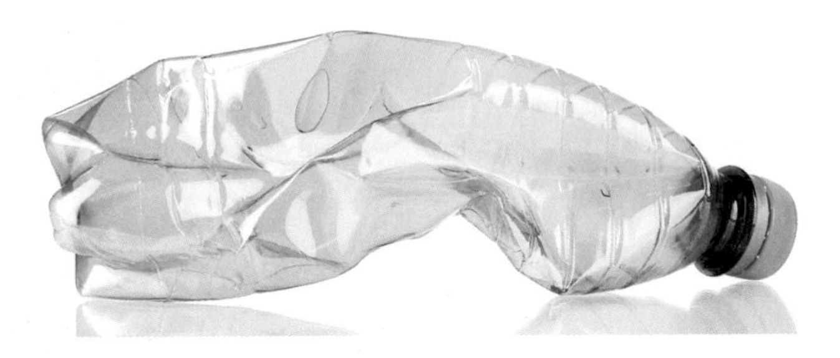

What Causes Air Pressure?

Air particles have mass, so they are pulled to Earth by gravity. The air surrounding Earth has weight. Atmospheric pressure is the weight of the air pushing on Earth's surface and on everything near Earth's surface.

Remember, air particles are zipping around individually. So what prevents gravity from attracting them all to the ground? Why aren't we walking around knee deep in a soup of oxygen and nitrogen particles?

The answer is **kinetic energy**. Kinetic energy is energy of motion. The gas particles have so much energy of motion that they are bouncing each other around in all directions. They resist being compressed by this constant banging into one another, at an astounding rate of billions of times per second.

Here's an important point: The atmospheric pressure is not only pushing down on Earth. The particles are also banging around and colliding with the air particles above and from the sides. As a result, atmospheric pressure acts with equal force in every direction.

But atmospheric pressure is not the same everywhere. Air pressure is caused by the mass of air being pulled to Earth. So what happens if you go up into the atmosphere, high above Earth's surface, where there is more air below you and less air above you?

If you have the good luck to go for a hot-air-balloon ride, you might find yourself 2 kilometers (km) above the land. Up there, 2 km of air is below you, so that 2 km of air is not applying pressure up where you are. Therefore, the atmospheric pressure is less at that altitude.

Because we live at the bottom of a sea of compressible air, the atmosphere is most dense at Earth's surface. At the surface, with the greatest amount of air overhead, we have the greatest pressure. The air particles are squeezed close together because gases can be compressed. And when more particles are present in a given volume, the gas is more dense. The air becomes less and less dense as you go higher in the atmosphere.

As you go up in the atmosphere, pressure decreases, the air expands (less force pushing the particles together), and the air gets less dense. Mount Everest is over 8 km tall. Up on top, the atmospheric pressure is only one-third the pressure and **density** it is at sea level. Have you ever seen pictures of climbers laboring up the highest reaches of the mountain? Most of them are using oxygen supplies. Why? Because there is only one-third as much air in each breath at that elevation, so only one-third as much oxygen. It takes an exceptional climber to reach the summit without extra oxygen.

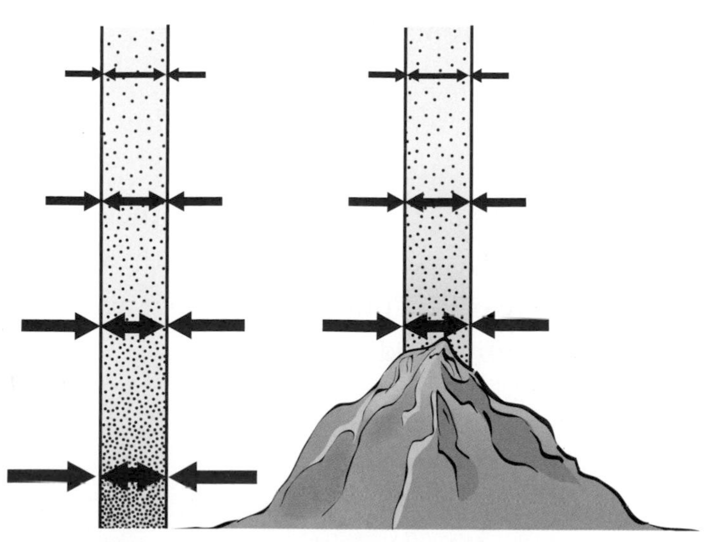
Air pressure on top of a mountain is less than at sea level.

Now, about those ears popping when you travel through the mountains. Does it have anything to do with atmospheric pressure? First, let's find out how atmospheric pressure is measured.

Measuring Air Pressure

The air pressure that meteorologists talk about on the evening news is caused by the mass of air pushing down on a certain point on Earth's surface. Elevation is one factor that causes pressure to vary, but there are a number of other factors as well. These factors, and the resulting pressure, are of interest to weather forecasters.

Meteorologists use a barometer to measure air pressure. An Italian naturalist named Evangelista Torricelli (1608–1647) invented the first barometer in 1643. He filled a long glass tube with mercury and turned it upside down in a dish also filled with mercury. A small amount of the mercury (not all of it) ran out of the tube and into the dish, leaving an empty space above the mercury. This space was a vacuum. A vacuum is a space containing almost no **matter**, not even air.

What was holding the heavy column of mercury up in the tube? Atmospheric pressure. Air pressure pushes down on the mercury in the dish. The pressure is distributed throughout the mercury, including the mercury in the tube. Remember, a column of atmosphere with an area of 1 cm^2 has a mass of 1.2 kg. If the column of mercury is 1 cm^2 in cross section, it will have a mass of . . . that's right, 1.2 kg. So, the force from air pressure is exactly balanced by the weight of the mercury.

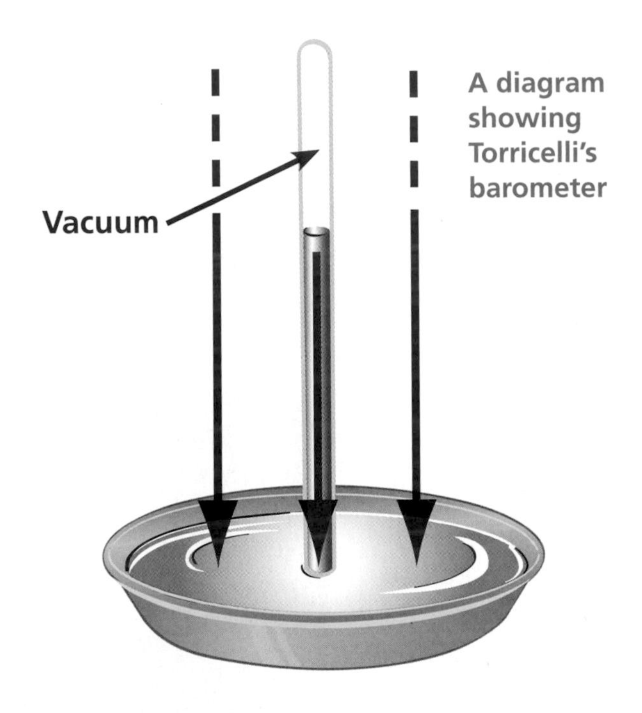

A diagram showing Torricelli's barometer

As Torricelli observed his new invention closely, he noted that the level of mercury moved up and down a little from day to day. He reasoned that the changing level of mercury was due to changes in the atmospheric pressure. If atmospheric pressure increases, there is more pressure on the mercury, so it rises up farther in the tube. Torricelli had invented the first barometer, an instrument for observing and measuring changes in atmospheric pressure.

Review your *Pressure in a Jar* notebook sheet. How is Torricelli's open-system barometer different from the closed-system pressure jar you used in class? How are they similar?

Today, meteorologists often use another type of barometer called an aneroid barometer. Aneroid barometers are much smaller and more portable than mercury barometers.

At the heart of an aneroid barometer is a sealed bellows-like chamber with a spring inside. All the air is removed from inside the bellows. As atmospheric pressure increases, air particles bump into the bellows more frequently. This compresses the bellows, and the device becomes shorter. On the other hand, lower pressure means less frequent hits from air particles, which allows the bellows to expand. A pointer attached to the bellows moves along a scale to show the change in pressure.

Modern barometers are digital. They use electronic pressure sensors in circuits that produce digital displays of atmospheric pressure.

If an electronic sensor and a transmitter are attached to a barometer, air pressure information can be radioed from a weather balloon back to a receiving unit on Earth. That's how we collect information about air pressure at high altitudes.

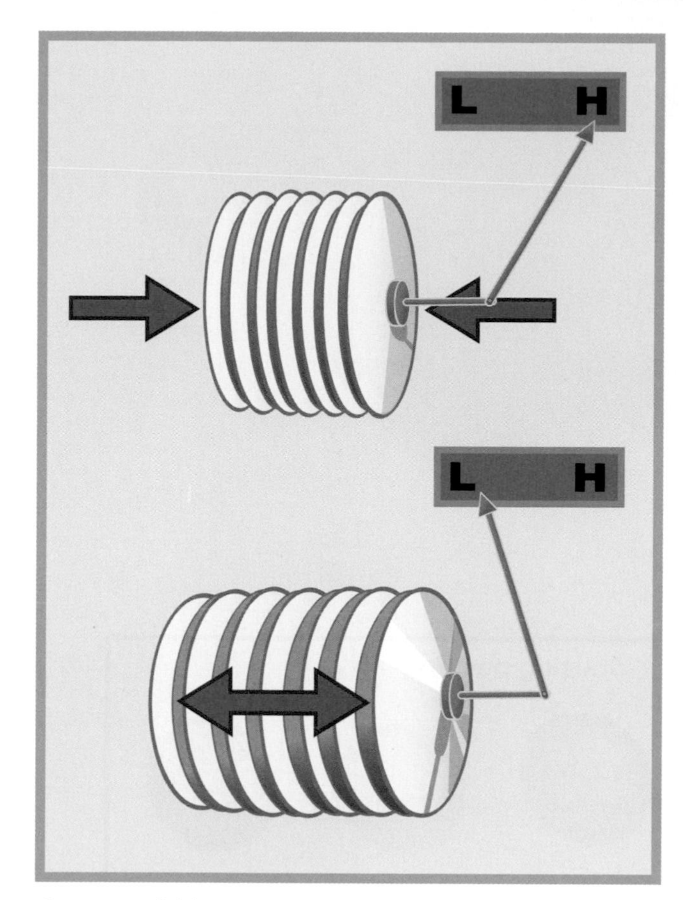

An aneroid barometer

When you packed your suitcase at home and included a half-filled bottle of lotion, the air pressure inside the bottle was equal to the air pressure on the ground. As it rides in the airplane up to the top of the troposphere, 80 percent of the atmosphere is now below, so the pressure inside the bottle is greater than outside. When you open the bottle, air will rush out and can force lotion along with it! A bottle with air in it is actually a simple barometer.

Review your work with the syringe. What might happen to air in a syringe if you took a clamped-shut syringe of air on an airplane?

Scientists and meteorologists use several different units to report pressure. For historical reasons, inches or centimeters of mercury are used in science experiments. Meteorologists, on the other hand, look at the average atmospheric pressure at sea level and call that 1 bar. If you are relaxing at the beach (sea level), the pressure around you will be 1 bar (or close to it).

The bar has been subdivided into 1,000 equal parts called millibars (mb). These are the units you used to record your local atmospheric pressure on the class weather chart. In practice, standard pressure is actually 1,013 mb. Any pressure below 1,013 mb is lower than normal pressure, and over 1,013 mb is higher than normal pressure.

Think Questions

1. When you drive down a mountain, what makes your ears experience those interesting and sometimes uncomfortable sensations?
2. Why doesn't air pressure crush an empty soft-drink can as you drive down a mountain?
3. Why doesn't air pressure crush an empty soft-drink can that is sitting on a table in your house?
4. If a meteorologist says that the air pressure is getting lower, what would you expect to see happen to Torricelli's mercury barometer?
5. If Torricelli had drilled a little hole at the top of the glass tube holding his mercury column, what would have happened to his barometer?

Density

Imagine you have a package of regular rubber balloons. Fill one with water, tie it off, and give it to a friend. Fill a second, identical balloon with air until it is the same size as the water balloon. Tie it off and give it to a second friend to hold. Fill a third balloon with helium, same size as the other two, and tie it off.

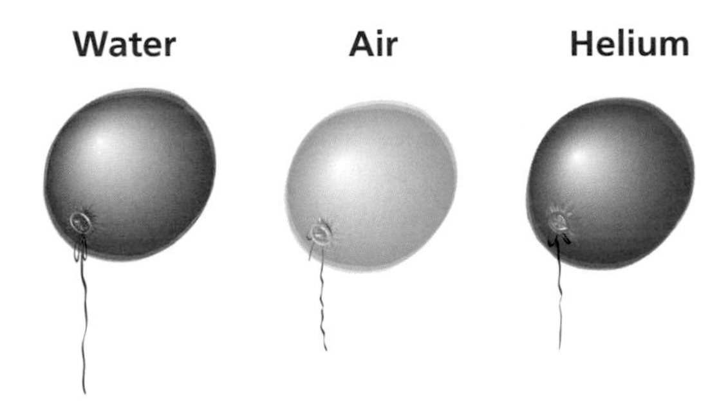

Review the three identical balloons, all filled to exactly the same volume, each tied off so nothing can get in or out. What's different? The kind of material in the three balloons. Ready to try a little experiment?

You and your friends hold the balloons at the same height above the floor. On the count of three, you will all release your balloons and observe what happens.

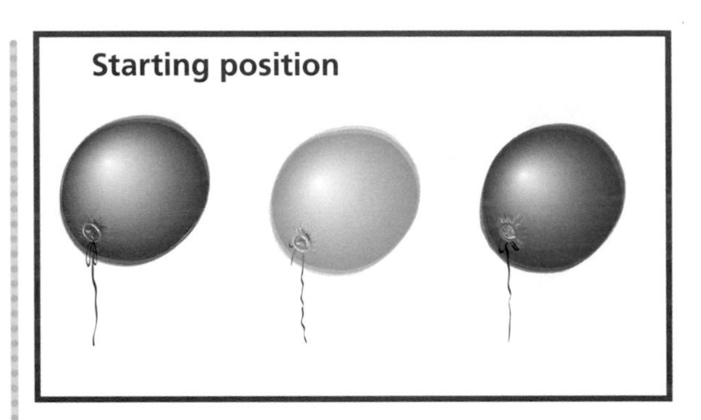

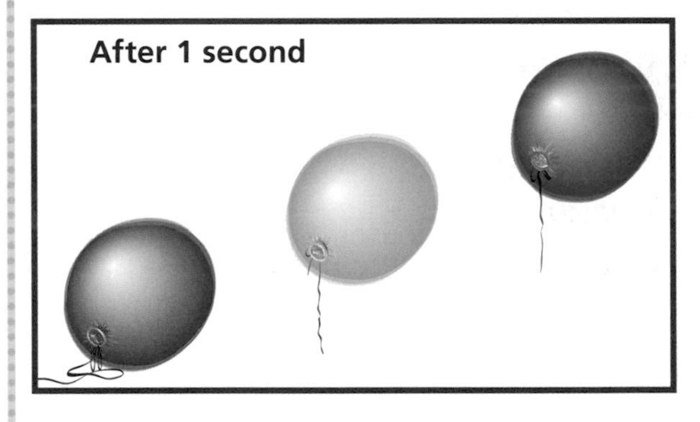

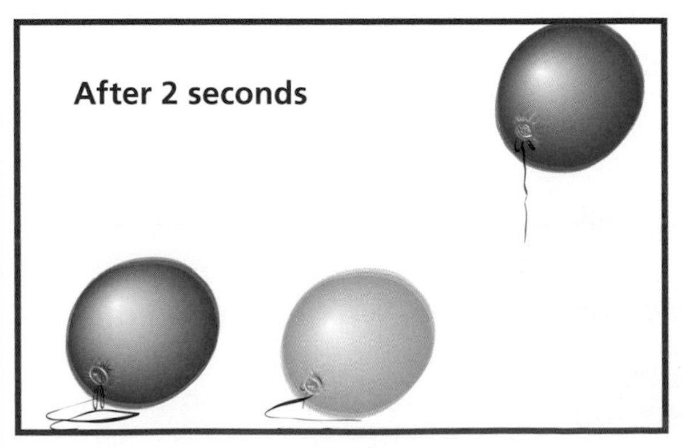

The water balloon will plunge to the floor, the air balloon will drift slowly to the floor, and the helium balloon will float up to the ceiling. Why?

It comes down to how much stuff there is in each balloon, relative to the stuff outside. The scientific word for stuff is matter. The amount of matter in an object is its mass. Matter is made of particles called atoms. So the mass of an object depends on *how many* atoms there are in the object and how big the atoms are.

The atoms in solids (glass, steel) and liquids (water, alcohol) are packed together as close as they can get. This means there are lots of atoms in a volume of water. That makes water pretty heavy.

In gases, the atoms are not packed as close together as they can get. There is a lot of space between atoms. Remember how you could compress air in the syringe? Air and helium are gases, so they are pretty light.

The nitrogen and oxygen atoms in the air are fairly large, but helium atoms are small. Generally speaking, small atoms weigh less than large atoms. So a volume of helium is much lighter than an equal volume of air.

Calculating Density

The amount of matter in a volume of material determines the material's density. Density is defined as mass per volume of an object. When you have equal volumes of a number of different materials, you can find out which one is most dense and which one is least dense by weighing them. The heaviest one is the most dense. The lightest one is the least dense.

The important idea in this discussion is that you need to compare the mass of *equal volumes* of different materials to determine which one is most dense.

The usual way of stating density is *mass per volume.* The word *per* means "divided by."

Density can be written as an equation.

$$\text{Density} = \frac{\text{Mass}}{\text{Volume}}$$

Density equals mass divided by volume. If you remember "mass per volume," you will always know how to set up your equation when it comes time to calculate density.

Density Interactions

One way to determine the density of a material is to determine the mass of 1 cubic centimeter (cm^3) of that material. If there is a lot of matter in a cubic centimeter of a material, it is dense. If there is very little matter in a cubic centimeter of a material, it is not very dense.

One milliliter (mL) is exactly the same volume as 1 cm^3. Liquids are generally measured in milliliters, and solids and gases are measured in cubic centimeters.

Mass is measured in grams (g). Pure water has a density of 1 g/mL. Materials with densities greater than 1 g/mL are more dense than water. Materials with densities less than 1 g/mL are less dense than water.

What will happen if you put a rock with a density of 3 g/cm^3 in a tub of water? It will sink like, well, a rock. What about if you put a cork with a density of 0.45 g/cm^3 in the tub of water? Yes, it will float.

Materials that are more dense than water sink in water. Materials that are less dense than water float in water. That's the way it always works.

Density of Gases

Back to the balloons. We know the density of water, but what about the air and the helium?

Material	Density
Water	1.0 g/mL
Air	0.0013 g/mL
Helium	0.0002 g/mL

The table shows that air and helium are not very dense. There is very little mass in a milliliter of gas. When the three balloons were released, the dense water-filled balloon pushed through the less-dense air surrounding it. It fell straight to the floor.

The air-filled balloon was almost the same density as the surrounding air. The rubber-balloon membrane is more dense than air, and the air was compressed a little inside the balloon, making the air-filled balloon a little more dense than the surrounding air. It drifted slowly to the floor.

The helium-filled balloon was quite a bit less dense than the surrounding air. Just like a cork in water, the less-dense helium balloon floated as high as it could.

Density of Air

Air is gas. The particles in gases are not bonded to other particles. Gas particles move around freely in space.

When energy transfers to matter, the kinetic energy (movement) of the particles increases. The increased motion causes most matter to expand. When matter expands, the particles do not get bigger. They get farther apart. This is a very important point: it is the distance between particles that increases, not the size of the particles.

When matter expands, the particles get farther apart. What do you think that does to the density of the material? The density gets lower. When the particles get farther apart, each cubic centimeter has fewer particles. Fewer particles in a volume means lower density.

This is a general rule of matter. When matter gets hot, it expands, and the density goes down. The idea of density will be an important concept in our investigation of weather and the processes that cause it.

Think Questions

1. What do you think the density of a person might be? Explain.
2. Why do you think hot-air balloons can rise into the air? How do hot-air-balloon pilots get their balloons back to Earth?
3. Weather happens in the atmosphere. What do you think happens when some air heats up and other air is cool? What weather might be the result?

Density with Dey

Mr. Dey's students had been working with salt solutions. When salt dissolves in water, it forms a solution. The students found out that the more salt they dissolved in a volume of water, the more dense the solution became. Mr. Dey's students received three salt solutions from another class and were challenged to order them by density.

Group 1 put 25 milliliters (mL) of the blue solution on a scale and found that it had a mass of 25 grams (g). They measured 25 mL of green solution into another cup. Its mass was 30 g.

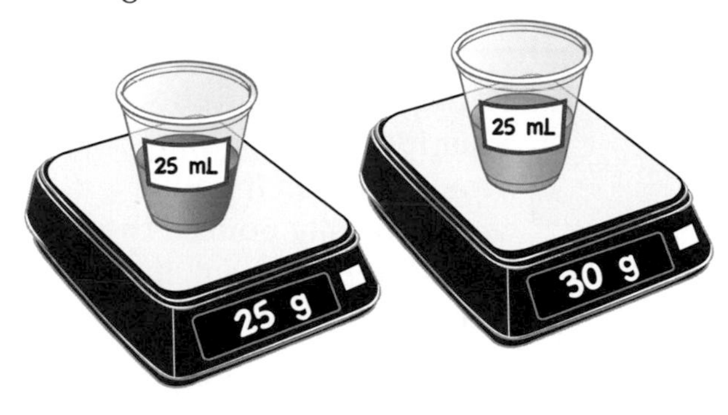

The students announced, "We weighed equal volumes of two solutions. The green solution is heavier, so it is more dense. It has more mass per volume than the blue solution."

Group 2 put 25 mL of blue solution on a scale and found that it had a mass of 25 g. But then they made a little mistake. They put 50 mL of yellow solution in a cup and found that it had a mass of 55 g.

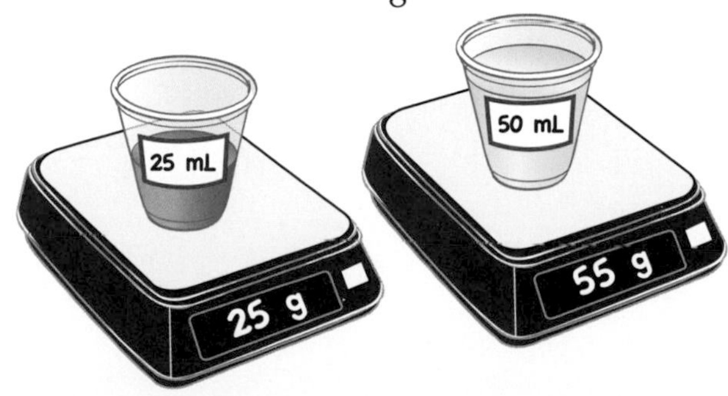

When they realized what they had done, Reggie said, "Uh-oh, we didn't measure equal volumes. We have to start over."

"Maybe not," said Yolanda. "We weighed twice as much yellow solution as we should have. If we had used half as much, it would have weighed half as much. All we have to do is divide the mass by two to find out the mass of 25 mL of yellow solution."

They did the math and found that 25 mL of yellow solution had a mass of 27.5 g.

The two groups put their data into a table.

Solution	Volume	Mass
Blue	25 mL	25 g
Green	25 mL	30 g
Yellow	25 mL	27.5 g

Students could now easily compare equal volumes of the three solutions to see which one was heaviest and, therefore, the most dense. They determined that green was most dense, blue least dense, and yellow in the middle.

Mr. Dey had a question. "What is the mass of 1 mL of each of the solutions?"

Reggie offered, "Twenty-five mL of blue solution has a mass of 25 g, so 1 mL of blue has a mass of 1 g."

"And the green solution?" asked Mr. Dey.

Reggie's group thought about it this way.

- Twenty-five mL of green has a mass of 30 g. That's more than 1 g for each milliliter.
- They drew a pie chart to help them think about the problem. Each slice of the pie represented 1 mL, or 1/25 of the total volume.

- One mL is 1/25 of the total volume. Each milliliter must have 1/25 of the total mass.
- They divided 30 g by 25 mL to find the mass of 1 mL of green solution.

The students discovered the definition of density.

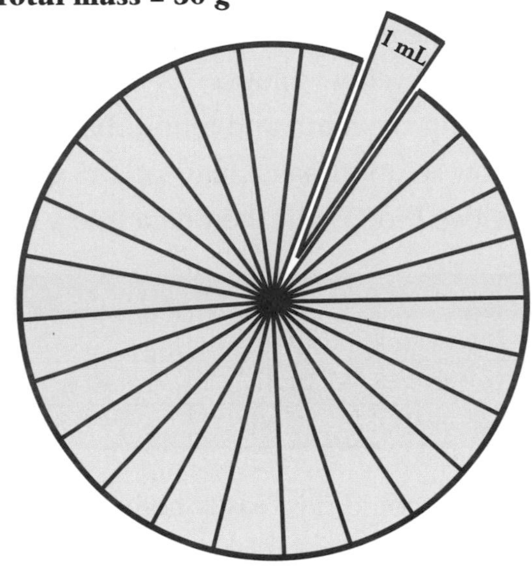

Density can be written as an equation.

$$\text{Density} = \frac{\text{Mass}}{\text{Volume}} = \frac{\text{g}}{\text{mL}}$$

The equation can be used to calculate the density of the green solution.

$$\text{Density} = \frac{\text{Mass}}{\text{Volume}} = \frac{30\text{ g}}{25\text{ mL}} = 1.2\text{ g/mL}$$

Density equals mass divided by volume. If you remember "mass per volume," you will always know how to set up your equation when it comes time to calculate a density.

What's the density of the yellow solution? Remember, mass per volume. The original mass of the yellow solution was 55 g, and the volume was 50 mL.

$$\text{Density} = \frac{\text{Mass}}{\text{Volume}} = \frac{55\text{ g}}{50\text{ mL}} = 1.1\text{ g/mL}$$

If you carefully pour a little bit of each of the colored solutions from Mr. Dey's class into a vial, what do you think would happen? Can you describe the result?

Solution	Density
Blue	1.0 g/mL
Yellow	1.1 g/mL
Green	1.2 g/mL

Record in your notebook your ideas about what would happen if you very carefully poured the solutions into a vial. Include a diagram to help you explain.

Convection

If you put a couple of centimeters of water in a metal pie pan and support it over a candle, you can slowly heat the water. The energy from the flame will heat a small area of the pan and the water directly over the flame.

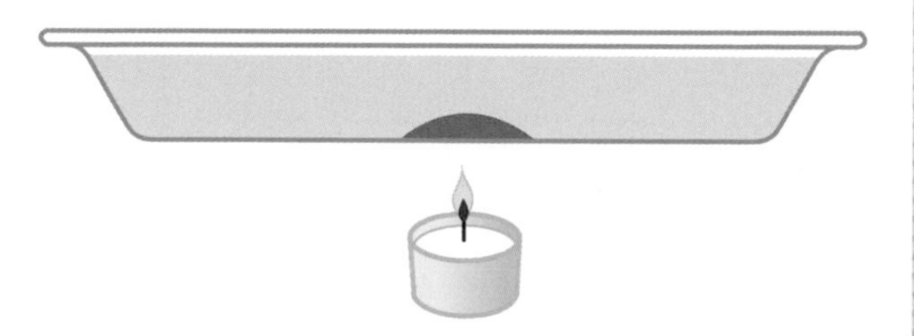

A small mass of water will heat up. The question is, what happens to the hot water?

The water expands as the water particles gain kinetic energy and push farther apart. The expanding water is less dense than the surrounding water. The warm water rises.

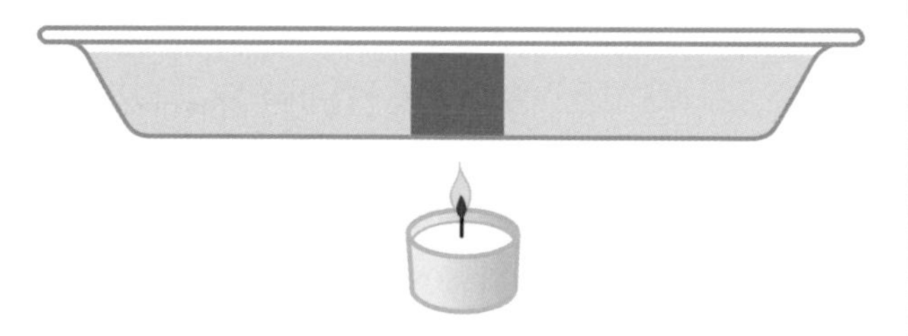

When the warm water reaches the surface, it spreads out. Water at the surface cools because it is no longer near the heat source and energy can transfer away from water to the environment.

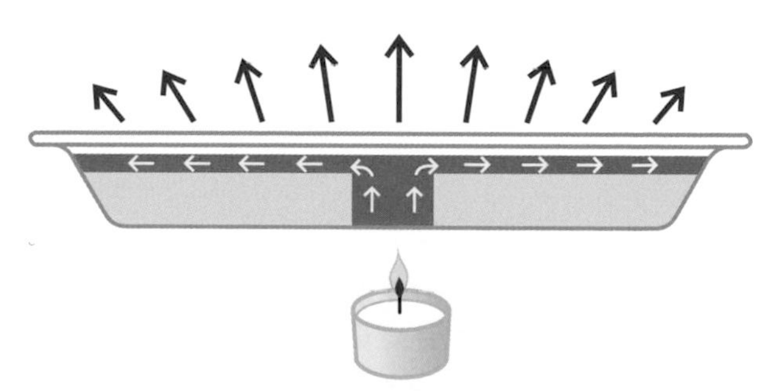

When the cool water reaches the edges of the pie pan, the water is dense. It flows down the sides of the pan, across the bottom, and back toward the center of the pan. As the water nears the hot metal, it begins to warm again. The hot water rises to repeat the cycle.

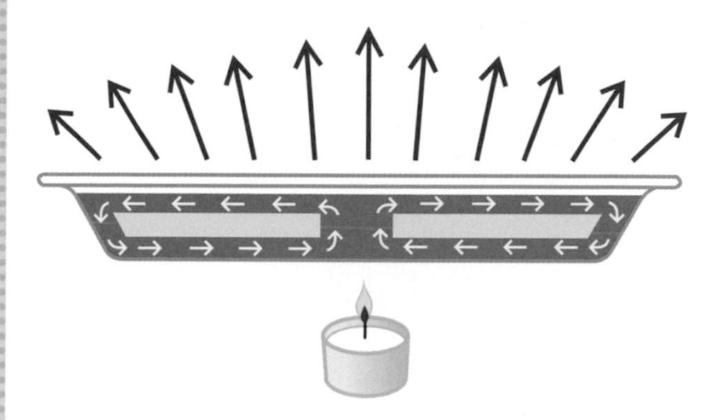

The movement of water in the pan, driven by a localized heat source, is **convection**.

Convection happens only in **fluids**. Gases and liquids are both fluids because they flow. Fluid near an energy source heats up and expands. The hot fluid becomes less dense and rises. The energy in the particles of hot fluid is carried to a new location. As the fluid cools down (heat energy transfers away from the fluid), it cools and contracts, making it more dense. The cool fluid flows downward again. The mass of fluid flowing in a circle is called a **convection cell**.

Air is a fluid. Air can heat up, causing it to expand. When air expands, it becomes less dense. Less-dense air will rise in the atmosphere. In the next few investigations, we will see how the process of convection helps redistribute water around the planet and affects all types of wind, from gentle breezes to powerful, dangerous storms.

Seasons

What do you imagine when you read these words: summer, spring, fall, winter?

Most of us come up with a mental picture or two. Summer means shorts and T-shirts, swimming, and fresh fruits and vegetables. Winter means heavy coats and short days, perhaps with a blanket of snow on everything. Seasons are pretty easy to tell apart in most parts of the United States. The amount of daylight, the average temperature, and the behaviors of plants and animals are a few familiar indicators of the season. But what causes the predictable change of season? What have you learned in class that helps you explain the reasons for the seasons?

As Earth Tilts

Let's start with a quick review of some basic information about Earth.

Earth spins on an imaginary axle called an **axis**. The axis passes through the North and South Poles. This spinning on an axis is called rotation. It takes 24 hours for Earth to make one complete rotation.

Earth travels around (**orbits**) the Sun. Traveling around something is called **revolution**. One revolution takes 365 and 1/4 days, which we call 1 year.

Make a notebook entry. Record the reasons for seasons on Earth. You can add more after reading this article, but record your first ideas now.

North Star

Earth isn't straight up and down on its axis as it revolves around the Sun. It is tilted at a 23.5° angle.

The average distance between the Sun and Earth is about 150 million kilometers (km). Earth's orbit is slightly oval, so Earth is sometimes farther away from and sometimes closer to the Sun. This distance is so insignificant that it is not related to the seasons.

It would seem logical that summer would be when Earth is closest to the Sun. That idea is wrong. Each year when Earth is closest to the Sun, the Northern Hemisphere experiences winter. The reasons for the seasons are linked to Earth's tilt, not the distance from the Sun.

Think about Earth orbiting the Sun. As Earth orbits, it also rotates on its axis, one rotation every 24 hours. Here's something important: Earth's North Pole points toward a reference star called the **North Star**. No matter where Earth is in its orbit around the Sun, the North Pole points toward the North Star, day and night, every day all year.

Is Earth closer to the Sun in winter or in summer? Is distance from the Sun a reason for seasons on Earth?

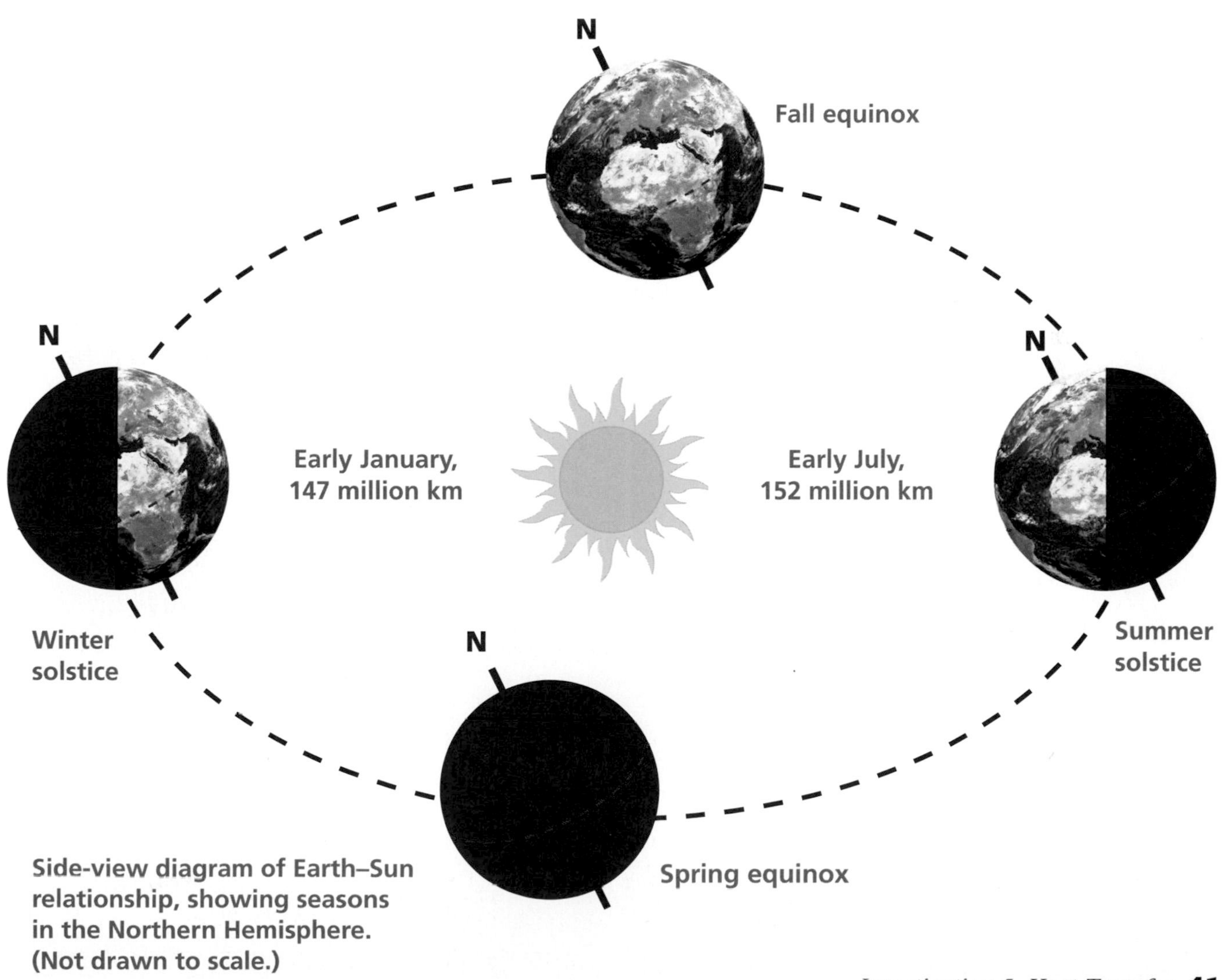

Side-view diagram of Earth–Sun relationship, showing seasons in the Northern Hemisphere. (Not drawn to scale.)

Tilt Equals Season

Look at the illustration on page 41. It shows where Earth is in its orbit around the Sun at each season. You will also see that the North Pole points toward the North Star in all four seasons.

Study the Earth diagram in the summer **solstice** position. Because of the tilt, the North Pole is "leaning" toward the Sun. When the North Pole is leaning toward the Sun, daylight is longer, and the angle at which light hits that part of Earth is more direct. Both of these factors result in more **solar energy** falling on the Northern Hemisphere. It is summer even though Earth is actually farther from the Sun. (And when it is summer in the Northern Hemisphere, it is winter in the Southern Hemisphere.)

Look at the position of Earth 6 months later (at winter solstice). Now the opposite is true. Even though Earth is closer to the Sun at this time, the Northern Hemisphere is tilted away from the Sun. Daylight hours are shorter, and sunlight does not come as directly to the Northern Hemisphere, so it gets less solar energy. It is winter in the Northern Hemisphere.

Four days in the year have names based on Earth's location around the Sun. In the Northern Hemisphere, summer solstice is June 21 or 22, when the North Pole tilts toward the Sun. Winter solstice is December 21 or 22, when the North Pole tilts away from the Sun.

The 2 days when the Sun's rays shine straight down on the equator are the **equinoxes**. On these 2 days, Earth's axis is tilted neither away from nor toward the Sun. *Equinox* means "equal night." Daylight and darkness are equal (or nearly equal) all over Earth. There are two equinoxes each year, the spring (also called vernal) equinox in March and the fall (also called autumnal) equinox in September.

Day and Night

We take day and night for granted. They always happen. Earth rotates on its axis, and the Sun appears to rise, then the Sun appears to set. This cycle has happened at least as long as humans have been on Earth. It will most likely continue for millions of years.

Because Earth is tilted, the length of day and night for any one place on Earth changes as the year passes. This table shows how hours of daylight change at different latitudes during the year. When it's summer in the Northern Hemisphere, the North Pole tilts toward the Sun. During this time at the North Pole, the Sun never sets. Above the Arctic Circle (66.5° north), daylight can last up to 24 hours of the day in the summer. Darkness can last up to 24 hours of the day during the winter.

Think Questions

Go back to the notebook entry about the reasons for the seasons that you made at the beginning of this article. What do you need to add? What do you need to change?

Length of Daylight in the Northern Hemisphere

Latitude	Summer solstice	Winter solstice	Equinoxes
0° N	12 hr.	12 hr.	12 hr.
10° N	12 hr. 35 min.	11 hr. 25 min.	12 hr.
20° N	13 hr. 12 min.	10 hr. 48 min.	12 hr.
30° N	13 hr. 56 min.	10 hr. 4 min.	12 hr.
40° N	14 hr. 52 min.	9 hr. 8 min.	12 hr.
50° N	16 hr. 18 min.	7 hr. 42 min.	12 hr.
60° N	18 hr. 27 min.	5 hr. 33 min.	12 hr.
70° N	24 hr.	0 hr.	12 hr.
80° N	24 hr.	0 hr.	12 hr.
90° N	24 hr.	0 hr.	12 hr.

Thermometer: A Device to Measure Temperature

Is the oven ready for this pie? You look flushed, do you have a fever? The fish are not eating. Is it warm enough in the aquarium? The ice cream is soft. Is that freezer working?

To answer these questions, we reach for a thermometer. And these days, there are lots of different kinds to reach for.

All thermometers work the same way on a basic level: some property of a material changes as it gets hot. You may already be familiar with several kinds of thermometers. The old standby is the glass tube filled with alcohol or mercury. This is how it works.

A thin, heat-tolerant, glass tube with a bulb at one end is partly filled with alcohol or mercury. The liquid extends partway up the tube. The tube is then sealed and attached to a backing that has a scale written on it.

When the bulb touches something hot, the liquid inside expands. The volume of liquid increases. The only place the added volume of liquid can go is up into the tube. The distance the liquid pushes up the tube indicates the temperature of the material touching the thermometer bulb.

The first closed-tube thermometer was invented by Ferdinando II de' Medici (1610–1670) in 1641. He used alcohol in the tube. During the 18th century, more precise closed-tube thermometers made it possible to conduct experiments involving fairly accurate temperature measurements.

In 1714, German physicist Daniel Gabriel Fahrenheit (1686–1736) made a mercury thermometer and developed the Fahrenheit temperature scale. On the Fahrenheit scale, 32 degrees Fahrenheit (°F) is the freezing point of water, and 212°F is the boiling point of water. In 1742, Swedish astronomer and physicist Anders Celsius (1701–1744) devised a temperature scale on which 0 degrees Celsius (°C) is the freezing point of water, and 100°C is the boiling point of water. This used to be called the centigrade scale, which means "hundred steps." But in 1948, it was renamed the Celsius scale in honor of Anders Celsius.

There are other types of thermometers. Oven thermometers and some wall thermometers look a little like pocket watches. Inside is a **bimetallic strip**. Bimetallic strips are made of two metals stuck together. The two metals expand at different rates when they get hot. When the heat is turned up, the copper-colored part of the strip expands (lengthens) more than the other. The strip bends. A pointer attached to the bending metal strip points to the temperature.

Tropical-fish enthusiasts keep thermometers right in the aquarium. One efficient kind is a thin, flat strip of plastic, like a piece of black plastic ribbon, that has liquid crystals packaged inside. Liquid crystals change color within a very narrow temperature range. A liquid-crystal thermometer has a series of little pockets in the strip, each filled with a different mix of liquid crystals to indicate one temperature only. So all you have to do is look at the strip to see which number is surrounded by a glow of color, and that's the temperature.

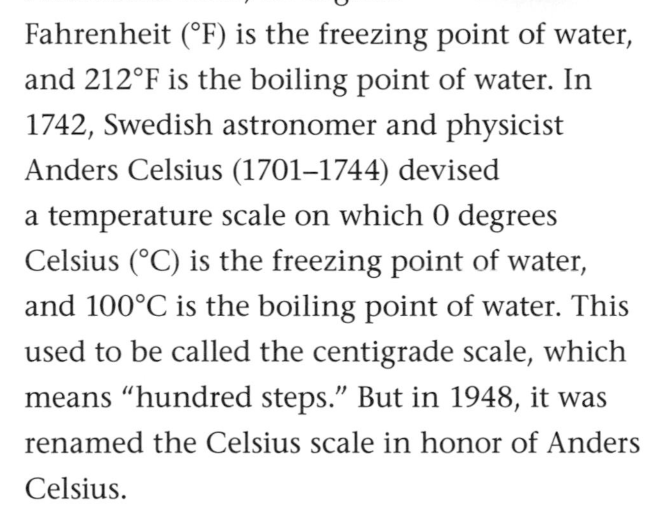

A bimetallic strip

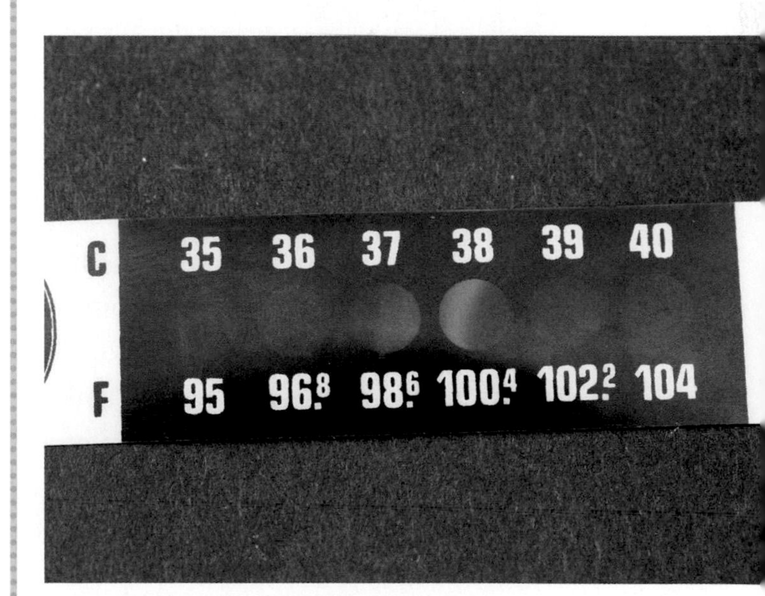

A liquid crystal thermometer

A digital thermometer

The last time you went to the doctor for a checkup, you may have had your temperature taken with a digital thermometer. These modern thermometers are very accurate and easy to use. You slip the probe end under your tongue for a few seconds. Inside the probe is a circuit with electricity flowing through it. Part of the circuit flows through a piece of wire that changes resistance as the temperature increases. When the electronic circuitry detects that the current flowing in the probe circuit has stabilized, that means the temperature is no longer changing. The electronic thermometer measures the amount of current flowing in the circuit and displays the temperature on a little digital screen.

That's just a small sampling of the many different thermometers found in common and specialized applications.

A Water-Density Thermometer

Galileo Galilei (1564–1642) invented one of the first functional thermometers in 1596. He filled a number of small glass balls partway with colored water and sealed them shut. The balls of colored water all floated in water because of the density of the glass ball systems.

Galileo knew that water expands as it warms up. Warm, expanded water is less dense than cold water. He then attached a weight to each ball. The weights were adjusted to give each ball a slightly different buoyancy. The result was that when the water was cold (at its most dense), all of the balls floated. As the water warmed up, becoming less dense, balls would sink.

By placing the balls in a column of water in order of their buoyancy, with the least dense on the top, Galileo produced a thermometer. If all of the balls were on the bottom of the cylinder, it was really hot.

Modern versions of the Galileo thermometer have temperatures printed on the weights. The number on the lowest floating ball shows the temperature of the system.

Heating the Atmosphere

The campfire has burned down to a bed of hot coals, perfect for toasting some marshmallows. You get a stick about half a meter long, since heat energy from the coals won't let you put your hand any closer. After a minute, the marshmallow is brown and gooey. You pop it into your mouth. Yikes! You didn't wait long enough for it to cool.

This story has two intense heat energy experiences that show two different ways energy can move: (1) the heat that you felt from a distance coming from the coals and (2) the heat that burned your tongue when you tried to eat the marshmallow.

What Is Heat Energy?

Objects in motion have energy that's called **kinetic energy**. The faster something moves, the more kinetic energy it has. Particles are always in motion, vibrating back and forth in solids and moving all over the place in liquids and gases. The amount of kinetic energy in the particles of a material determines its temperature. In hot materials, particles are moving faster, and in cold materials, they are moving more slowly.

Conduction

Heat naturally transfers from a hotter location to a cooler one. In our story, the heat in the hot marshmallow transferred to your tongue, a memorable experience that is an example of what happens when heat energy moves from one place to another by direct contact. That kind of energy transfer is called **conduction**. The faster moving particles of the marshmallow transfer energy to the slower moving particles of your tongue. If the particles of your tongue gain too much kinetic energy, you feel it as pain. At the same time, particles of the marshmallow lose kinetic energy, so the marshmallow cools off. Some other examples of conduction are when you take a hot bath and heat moves from water to you, or you jump into a cold lake and heat moves from your body to the water. Brrrr!

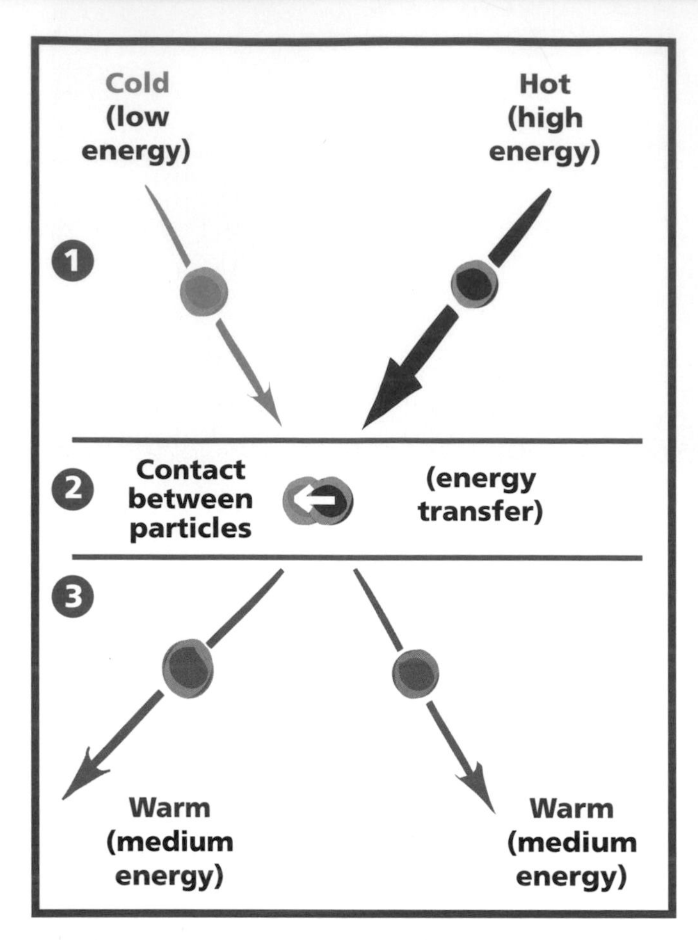

A "hot" particle with a lot of kinetic energy collides with a "cold" particle with little kinetic energy. Energy transfers at the point of contact. The cold particle then has more energy, and the hot particle has less energy.

Radiation

What about that heat that you felt from a distance coming off the coals in our story? That energy traveled right through air without any direct contact. That is called **radiant energy**. It can radiate, or travel in rays, from sources like the hot campfire coals, lightbulbs, and the Sun. Rays from the Sun travel right through the empty space between Earth and the Sun. Some of the Sun's rays are visible light rays, and some are invisible, such as **infrared** rays (beyond the red end of the rainbow) and ultraviolet rays (beyond the violet end of the rainbow).

The Sun is a source of radiant energy.

Record your ideas for each question in your notebook. What are two ways energy can transfer from one material to another? Can you think of other examples of heat conduction? Radiation?

Heating the Atmosphere

Our atmosphere heats up by the same two kinds of energy transfer we identified in the campfire story: radiation and conduction. Starting with radiation, the visible light rays from the Sun pass right through our atmosphere, strike Earth, get absorbed, and heat up the landmasses and ocean. These heated landmasses and ocean come in contact with air particles, and voilà, heat energy is transferred to the air by conduction.

Infrared rays from the Sun are a different story. Many of the infrared rays make it all the way through the atmosphere like visible rays to heat landmasses and the ocean, but some of the infrared rays get absorbed by certain gases in the atmosphere. Gases that can absorb infrared rays are called greenhouse gases. Carbon dioxide (CO_2) and water vapor (H_2O) are both greenhouse gases. They can absorb infrared rays and heat up.

The warm land and ocean also radiate heat energy in the form of infrared rays. This is a very important idea. Greenhouse gases absorb these infrared rays just as they absorb the infrared rays from the Sun.

Greenhouse gases have earned this nickname because of the way they "trap" heat in Earth's atmosphere. The greenhouse gases in the atmosphere, including water vapor and carbon dioxide, absorb infrared rays (heat up) and then reradiate the heat energy in any direction, including back toward Earth. This means if the infrared ray was headed away from Earth, perhaps after being radiated from a hot asphalt schoolyard, it has some chance of being absorbed and reradiated instead of leaving the atmosphere. The more greenhouse gases, the greater this chance.

Without any greenhouse gases, Earth would lose a lot more heat to space. Earth's average temperature would be about 33 degrees Celsius (°C) colder than it is. Compare this to the Moon, which has temperatures from 123°C to -233°C (average -55°C). Because there is no atmosphere, all the heat from the daytime radiates back into space during the night. Greenhouse gases cause Earth to remain at a stable temperature that permits life.

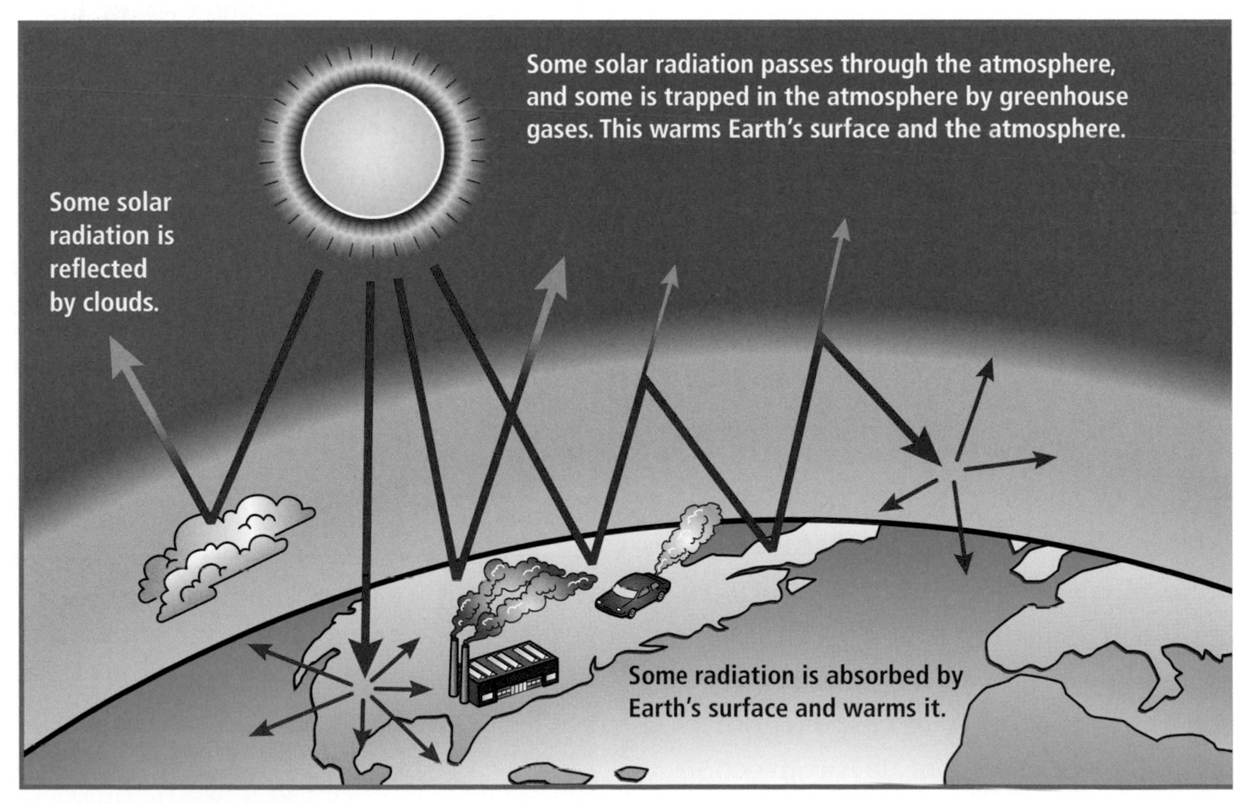

The greenhouse effect

So, the surface of Earth gets heated by both visible light rays and infrared rays through radiation. In addition, greenhouse-gas particles in the atmosphere get heated when they absorb infrared rays through radiation, either directly from the Sun or from a heated surface on Earth. Radiation won't warm up nitrogen and oxygen particles, which make up most of the atmosphere and are not greenhouse gases. But all gases in the atmosphere heat up when they come in contact with the heated surface of Earth, and energy transfers through conduction. In addition, the greenhouse gas particles in the atmosphere that were heated by infrared radiation will come into contact with nitrogen and oxygen particles and transfer heat energy to them through conduction.

The final piece of the puzzle is convection. Convection involves moving masses of air or water, both fluids. At the hottest parts of Earth's surface, huge masses of air heat up, expand, get less dense, and rise. This is the beginning of the largest convection cell on Earth. Over the ocean, large amounts of heat and water vapor rise, riding along in the giant convection cell. When the air cools, water vapor condenses into droplets of liquid water, forming clouds or rain. Convection helps redistribute heat and water around the planet, generating weather conditions ranging from mild to severe.

Record your ideas for each question in your notebook. Explain two ways that Earth's atmosphere gets heated. How do greenhouse gases help keep Earth warm?

Greenhouse Gases and Climate Change

The fact that greenhouse gases can help keep Earth warm has implications for humans and other organisms on Earth. If we change the amount of greenhouse gases, can we change the amount of heat in the atmosphere? One of the greenhouse gases humans regularly produce is carbon dioxide. It's a by-product that is released every time we turn on a gasoline-powered car or burn coal to generate electricity, among other sources. If you imagine all the electricity and gas used by each human in industrialized nations every day and multiply that by the number of humans . . . well, you can imagine why the amount of carbon dioxide in the atmosphere is increasing. Scientists have tracked carbon dioxide in the atmosphere since the 1950s, as shown in the graph.

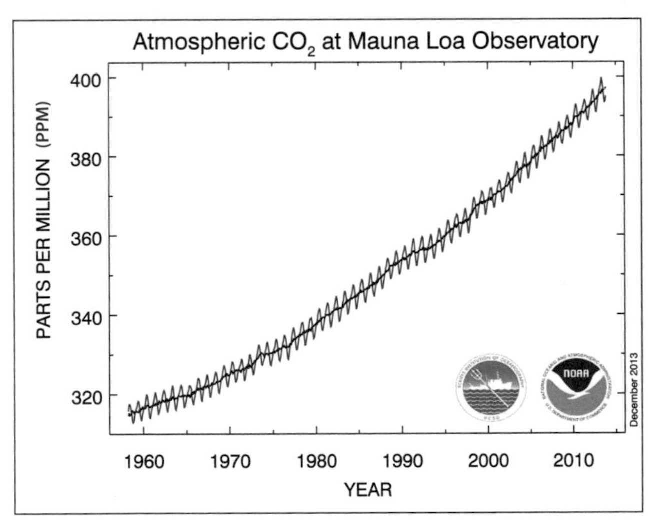

Scripps Institution of Oceanography
NOAA Earth System Research Laboratory

Think Questions

1. What patterns do you see in the data for atmospheric carbon dioxide?
2. What might happen on Earth if humans increase the amount of greenhouse gases in the atmosphere?

Laura's Big Day

Laura opened her eyes with a start. She checked her alarm clock. Only 6:30 a.m. Laura got out of bed and went to the window. All she could see was the dim outline of the tree in the front yard. Fog blocked out everything else on her street.

It was her birthday. How could the weather be so lousy? Laura really wanted the fog to go away. After all, she had something very special to do today. Uncle Ken was taking her hang gliding for her 13th birthday.

They had begun planning the event 6 months ago, about the time Laura reached 5 feet tall. Ever since she could remember, Laura had wanted to fly like a bird. She had stood on the tops of rocks with her arms stretched out, while her parents called to her to come down. She had flown in an airplane when she visited her grandparents, but that just wasn't the same. Laura peered into the fog, hoping to see the mountains beyond.

Laura was dejected and wide awake. When she went into the kitchen, she found her mom making coffee. "The weather is terrible!" Laura pouted.

"It will probably clear up by noon," her mother offered. "I saw the weather forecast last night. The meteorologist said that it was really hot out in the desert yesterday. A high-pressure area pushed ocean air over our town all day yesterday. When the humid sea cools off at night, the water vapor condenses, forming fog. I think the Sun will come up and warm the ground and the air. The fog will evaporate again, leaving us with a nice day."

Laura wasn't convinced. She ate her cereal and stared out the window.

"I'm going to the nursery later. Want to come along?" her mother asked.

Laura was tempted, but decided she would wait to see what the weather would do. "No, thanks," she replied.

At 9:00 a.m. it started getting lighter. Laura called Uncle Ken on the phone.

"Hi, Laura," Uncle Ken said. "Are you looking forward to your lesson today?"

Laura could hardly believe her ears. "Are we really going to be able to fly?" she blurted out.

"Sure," Uncle Ken said. "It's really gorgeous up here on the mountain, and I can see that the fog down your way is burning off."

"What do you mean, burning off?" Laura asked.

"The Sun warms up the air and the fog evaporates," Uncle Ken replied.

"Oh," said Laura, "that's just what Mom said."

"Get your mom to drive you up here, and we'll get the equipment ready. I'm already working on our flight plan."

Uncle Ken lived in a cabin on a mountain ranch where he worked when he wasn't teaching hang gliding. When Laura arrived at the cabin, she trotted down to the storage garage where Uncle Ken was working.

"What are we going to do?" she asked.

"Well, your first lesson will be a tandem flight so I can show you what hang gliding is all about," Uncle Ken answered. "We'll fly together on a glider. It's packaged for the road on top of my truck. I have to check on the weather now."

Uncle Ken had taken Laura with him in the spring when he taught hang gliding. She knew the names of the parts of the glider and had tried to balance one on her shoulders. She had listened to Uncle Ken talk about mature decisions and safety. Laura already knew about helmets, reserve parachutes, and harnesses. She had been at the landing zone with Uncle Ken, watching his friends land their gliders after long flights riding the thermals. She saw them look at the wind sock for wind direction, so they could land into the wind.

Uncle Ken had explained that summer is a good time for thermals. A thermal is a bubble or column of warm, rising air,

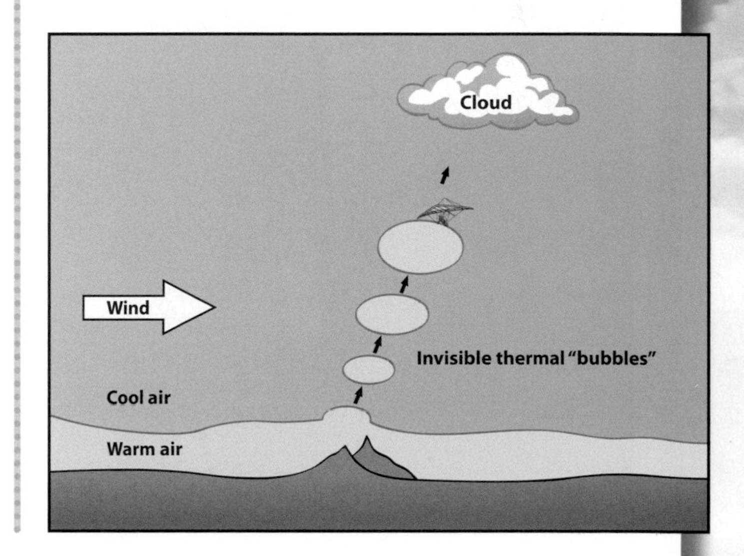

common on sunny days. The Sun heats the ground and the air near the ground. The air expands, making it less dense than the air above it. The warm air rises, like bubbles rising in a soft drink. An important skill in flying a hang glider is to spend as much time as possible in the strongest part of the thermal (usually the center).

Laura had seen Uncle Ken's weather station behind his cabin and how it was hooked up to an old computer. He recorded wind speed, wind direction, temperature, humidity, and atmospheric pressure. She also knew that he got data about current temperature and wind speed from a weather station near the landing zone.

A solar-powered weather station

Uncle Ken used the temperature difference between his cabin on the mountain and the landing zone to figure out how strong the thermals would be. As you go higher in the atmosphere, the air cools off. It cools off at a rate of about 3 degrees Celsius (°C) per 305 meters (m). Because the difference from the top to the bottom of the mountain was over 1,500 m, the air should cool by about 15°C. If the difference in temperature was more than that, it meant good thermals, because the warmer air at the bottom rises more quickly through the cooler air near the top of the mountain.

"Conditions look great. Let's go!"

Laura tossed her duffle into the back of Uncle Ken's truck. She and her mom jumped in, and they were off.

On the road, Laura asked, "Uncle Ken, how long will we be up in the air?"

"No telling," he responded. "I was up over 2 hours on Wednesday, but I don't think we'll push it that far. Maybe we'll fly for 15 or 20 minutes. But it all depends on the wind."

Uncle Ken talked about how this site had both ridge and thermal lifts, and that promised a great ride. "Ridge lift means that the wind gets forced straight up when it runs into a cliff or hill. Hang gliders don't have motors to gain altitude, you know, so the only energy we use for going higher in the air is from currents going upward, called updrafts. Any time that a steady wind blows directly against a wide slope, ridge lift will probably result."

They pulled into the launch site and parked near several other trucks and vans. Uncle Ken pointed to a wind sock about half extended by the gentle, steady breeze. "It's a little old-fashioned, but we like it. It's a pretty straightforward tool for estimating wind direction and speed. Looks like we have a breeze of about 24 kilometers (km) per hour."

Uncle Ken set about assembling the glider. It was surprisingly big, but he handled the lightweight materials with ease. Laura watched with interest, but she could feel a little anxiety rising in the pit of her stomach as the assembly neared completion. The moment when they would launch into thin air was drawing uncomfortably close.

Laura zipped up her warm parka and strapped on her harness.

"Is it tight enough? Does it feel right?" asked Laura's mother.

"Mom, please, I'm all right. Uncle Ken showed me how . . . I can do it by myself."

Uncle Ken checked and double-checked Laura's preparation. She was dressed in warm, rugged clothes, helmet, goggles, and a properly tightened harness. Ready!

Uncle Ken and Laura reviewed the procedure for getting into the air safely, the flying position, and the landing procedure. Uncle Ken could see how excited she was, so he reminded her that this was a lesson, not a theme-park ride, and that she had to be mature and follow his instructions in the air.

Uncle Ken hooked onto the hang glider and carried it to the launch area. Following his instructions exactly, Laura hooked her harness onto the glider right next to her uncle. She practiced lying out prone, just as she would in the air.

"Ready?" asked Uncle Ken.

"I think so. It's a little scary, but I want to see what it's like to fly."

Laura stood next to Uncle Ken on the edge of the gentle slope. The breeze was blowing in their faces. Laura felt the hang glider lift up as it caught the wind.

"Now! Run!"

Down the slope they dashed. Amazingly, in just a moment Laura's feet barely touched the ground. She felt as light as a leaf on the autumn wind. Then she wasn't touching the ground at all. She was in the air, and the ground was moving away with alarming speed.

"Uncle Ken, it worked! We're flying!"

Uncle Ken shifted his weight to bank the glider so they swooped over the launch area and Laura's mother. Laura waved at the amazed figure of her mother with her hands clamped over her mouth. Then her mother waved back and remembered to snap a few pictures.

"Let's use this nice ridge lift to climb a little, and then let's look for a thermal." They flew back and forth along the ridge twice, and then Uncle Ken turned away from the ridge to look for thermals. It was so quiet and peaceful. The hang glider just floated. As they headed over the valley, the land fell away, and Laura realized she was hundreds of meters above the ground. This was real flying.

"Look, there's a thermal just to the left," announced Uncle Ken after a couple of minutes. "Can you see it?"

She looked where he indicated, but could see nothing.

"Those two red-tailed hawks, spiraling up, have pointed out the invisible thermal for us. Let's join them."

In a minute, Laura could feel the turbulence in the air and the definite lift provided by the thermal. Uncle Ken skillfully banked the glider to keep it in the thermal, and they spiraled up, just like the hawks above them.

In a few minutes, the thermal lift died out, and they glided off toward the site where they would land, making large, lazy circles in the sky. When they passed through a region of turbulence, Laura asked with wide eyes, "Uncle Ken, was that a thermal? Could it take us higher again?"

Pleased by her close observation, Uncle Ken responded, "I'm not sure. Let's go see."

He banked sharply and entered the thermal. The glider was once again lifted over a hundred meters in a minute.

"Excellent lift. That was a nice little elevator you spotted. Let's head for home."

Uncle Ken turned the glider away from the mountain and began to fly a circular route that took him lower and closer to the landing area. As they got closer to the landing zone, Uncle Ken asked Laura to watch the wind sock and the streamers to see which way the wind was blowing.

They slowly glided lower and lower until the ground was just below them. Uncle Ken pushed the down tube forward, and the glider made an easy dive toward the landing target. They skimmed just above the ground. Uncle Ken pushed out hard on the down tube. The front of the glider rose up and they settled easily to the ground.

It was over. They were back on the ground. She had flown. Laura was dazed. "Uncle Ken, that was great! I never knew anything could be so exciting. Can we go again?"

"Well, the first hang-gliding trip is a ride. If you go up again, you will have to do some of the driving, you know."

"Can I? Can I really do the flying?"

Laura saw her father coming across the field as she and Uncle Ken carried the marvelous flying machine to the parking lot.

"Dad, I was flying and it was the greatest thing. It's like flying a kite, except we were on the kite! And you won't believe it, Dad, we followed red-tailed hawks. We flew with red-tailed hawks right to the top of a thermal!"

Laura couldn't get the idea of flying with birds out of her mind. She knew that she would continue her lessons and become a pilot. She'd have her own glider one day. But before that, there was a lot to learn about seeing, feeling, and imagining what's going on in the air, interpreting the weather, and what makes a good day for hang gliding.

Think Questions

1. Why is the weather forecast so important for hang gliders?
2. How does a hang-glider pilot rise higher in the atmosphere?

Wind on Earth

Earlier, you observed the air-flow pattern in the convection chamber. Air at one end of the chamber was being heated by a candle. Cool, smoke-filled air entered the chamber at the other end and immediately sank to the bottom of the chamber. The cool air moved across the bottom of the chamber and moved under the warmer, less-dense air, causing the warm air to rise. The rising warm air had no place to go, so it traveled across the top of the chamber to the other end. Here the air cooled and sank. The whole cycle repeated, producing a small convection cell.

This is a model of what happens in the real world. You have seen that Earth materials heat and cool at different rates (remember the experiment with sand, soil, water, and air). You also built a model to show how this differential heating creates convection cells and produces wind.

Now let's put everything you have learned together and look at differential heating on a global scale. Because of the high solar angle, Earth's surface is always warmer in the tropics, the part of the planet near the equator. Water in the tropical ocean and tropical landmasses absorb a lot of solar energy. Air in contact with the tropical sea or land receives a lot of energy by radiation and conduction. Huge masses of air heat up and begin to rise. This is the start of the largest convection cell on Earth, one the size of the entire equatorial area.

Earth convection cells

The equatorial convection cells circle the globe like two bicycle inner tubes. Because much of the energy transfer occurs over the ocean, large amounts of water vapor rise high into the atmosphere, riding along in the convection cell.

As the warm, moist air rises, it cools. At about 10 kilometers (km) altitude, the warm air has cooled to the same temperature as the surrounding air. The cool air begins to fall back to Earth. But because the wall of warm air is rising from the tropics, it can't return directly. The cool air, still at about 10 km, flows north and south, like two gigantic conveyor belts. When it reaches about 30 degrees (°) north and 30° south, it descends toward Earth.

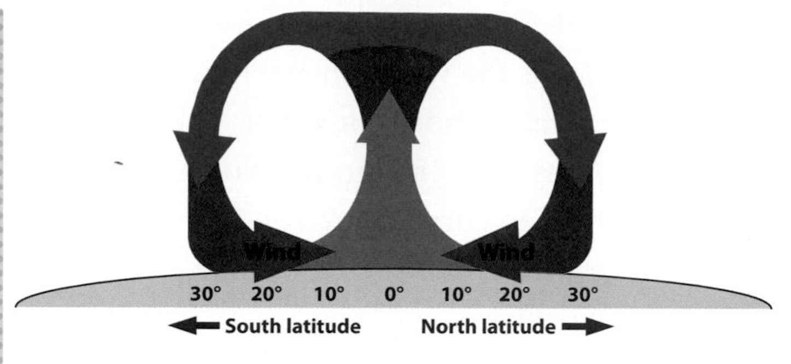

A generalized model of a convection cell in Earth's atmosphere

Meanwhile, the warm, low-density, rising air in the tropics creates a low-pressure area. The cooler, more-dense air that is descending from the upper atmosphere creates a high-pressure area. The more-dense air flows into the area of low pressure to replace the rising air. This creates a huge convection cell called a **Hadley cell**, named after George Hadley (1685–1768). He was the first to propose the idea of these enormous convection cells in 1735. The bottom of the cell flows across the surface of the planet, from about 30° north and south toward the equator, producing wind.

Prevailing Winds

Differential heating creates high- and low-pressure areas, causing wind. Winds always tend to move from a high-pressure area to a low-pressure area. However, as you saw earlier when you looked at air pressure maps and actual wind data, winds don't blow directly from high-pressure areas to low-pressure areas. Instead the wind takes a curving path. Why does this happen?

As air in the atmosphere moves, Earth rotates, dragging the air along with it and changing its direction. This is called the

Coriolis effect. This easterly movement of the planet means the winds in the Hadley cell don't go straight north or south. The Coriolis effect causes the winds in the Northern Hemisphere to curve clockwise and the winds in the Southern Hemisphere to curve counterclockwise.

The combination of high- and low-pressure areas and the Coriolis effect gives us the **prevailing wind** direction. Prevailing winds are global winds. They are predictable for different latitudes on Earth. These global winds are not greatly influenced by structures on Earth's surface, like forests, cities, and mountains. Prevailing wind direction is important to people sailing ships, flying balloons around Earth, flying airplanes, and building wind turbine farms to generate electricity. As you will see later, prevailing winds are also an important factor in the local weather.

Centuries ago, sailors named the different prevailing winds. Between 5° and 25° north or south latitude, there were usually dependable winds that would carry a ship across the ocean quickly. Because ships were crossing the ocean back and forth carrying goods to trade between the Americas and Europe, these winds became known as the **trade winds**.

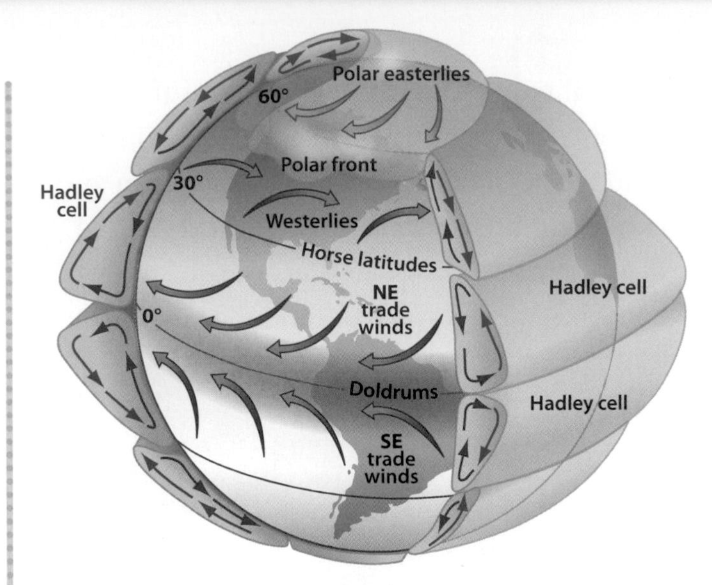

Prevailing winds on Earth

However, near the equator the air is rising, and at about 30° north and south latitudes, the air is descending. At these latitudes the wind is usually very light or nonexistent. Sailing ships crossing these areas could get stuck for days or weeks. The calm area around the equator became known as the **doldrums**. The windless area around 30° north was known as the **horse latitudes**. This term might have been coined when Europeans were carrying horses across the ocean to their colonies. Ships would sometimes get stuck at about 30° N, making the trip much longer than planned. The sailors would run out of water for the horses, and horses could die. Prevailing winds and Hadley cells happen at the surface of Earth.

There are winds at upper levels of the atmosphere as well, including the most familiar one called the **jet stream**. Jet streams are narrow bands of high-altitude wind that occur in predictable patterns, but vary over time. Changes to the jet streams can greatly affect weather patterns on the surface of Earth, and even change the flight time of an aircraft.

In addition to the Hadley cells, there are two other bands of convection cells in the Northern and Southern Hemispheres. They produce prevailing westerlies (wind blowing from the west) between 35° and 55° north and south latitude and polar easterlies (wind blowing from the east) north and south of 60°.

This tree shows evidence of growing in an area with strong prevailing winds.

Local Winds

Local winds change with the season and even with the time of day. They are the direct result of local differential heating. Local winds are affected by land structures and bodies of water near landmasses.

A typical Chicago weather forecast in the summer might go something like this: "Sunny skies today with a high of 29°C inland. Temperatures will be in the low 20s lakeside." The climate of Chicago is affected by being near a large body of water (Lake Michigan). Landmasses near the Great Lakes or the ocean will feel the effects of the nearby water. Here's how.

From your experiment with the heating of earth materials, you know that landmasses get hotter faster than water when the Sun shines. The hot land transfers heat to the air above it, and the air expands. The warmer, less-dense air creates a low-pressure area over the land.

Even though the sea also absorbs energy, its temperature does not change much at all. During the day, the sea stays cooler than the land. Less energy transfers to the air over the water, so it is cool and dense, which means higher air pressure. Air from the high-pressure area over the water flows into the area of low pressure. Wind blows from the sea onto the land. This is called a **sea breeze**.

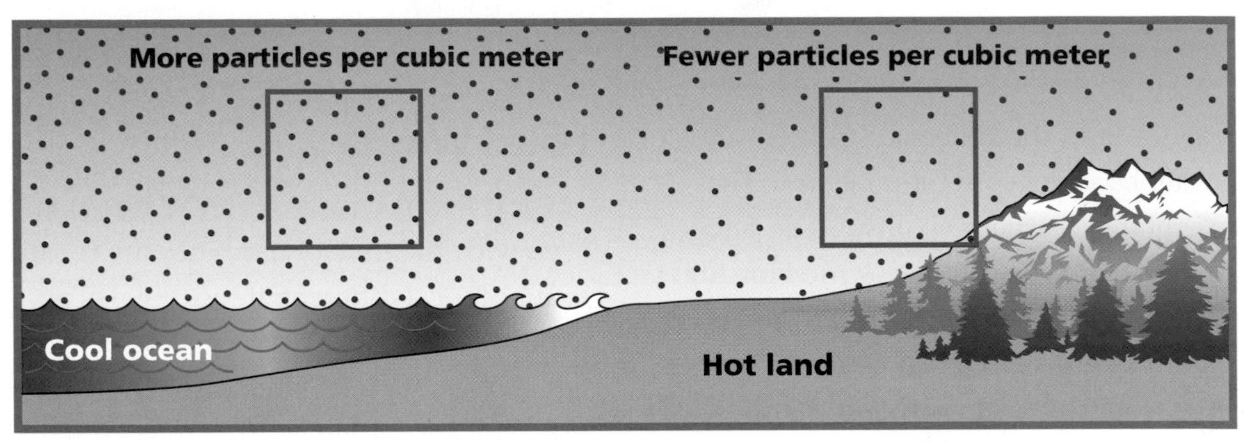

A cubic meter of hot air has fewer particles than a cubic meter of cold air. Hot air is less dense than cool air.

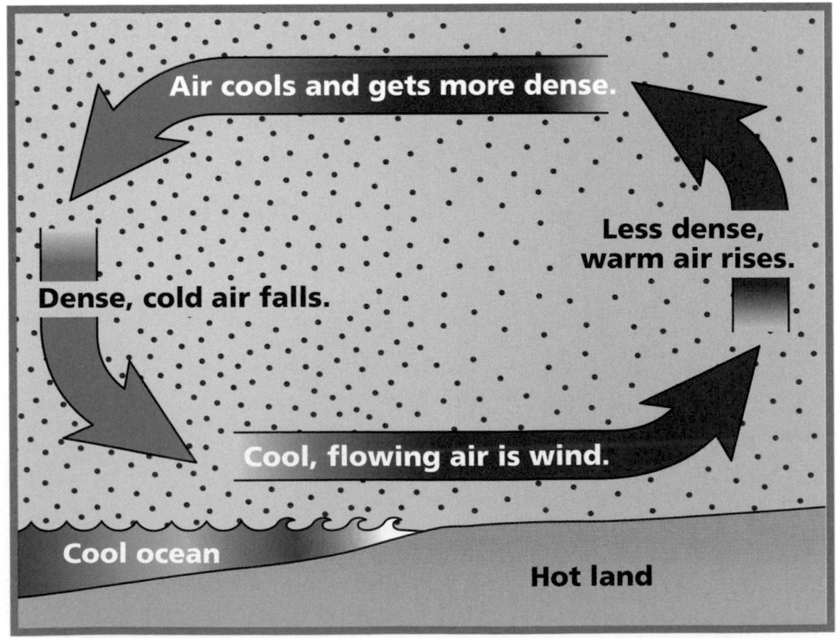

A convection current is the result of the uneven heating of Earth's surface.

At sunset, there may be a period of calm when land and sea temperatures are about equal. After sunset, the land cools off quickly. The air over the land cools and contracts, becoming more dense. At this point, the local high-pressure area is over the land. Meanwhile, the sea temperature is still about the same as it was during the day, so the air pressure is about the same, but it is lower than the local pressure over the land. Now the local high pressure over the land moves into the lower-pressure area over the water. Wind blows from the land out to sea. This is called a **land breeze**.

Think Questions

1. What causes the low-pressure region found throughout the tropics?
2. Why is there almost no wind at the equator and at 30° north latitude, but a dependable northeasterly wind in between?
3. What causes local winds to blow from the sea onto the land during the day?

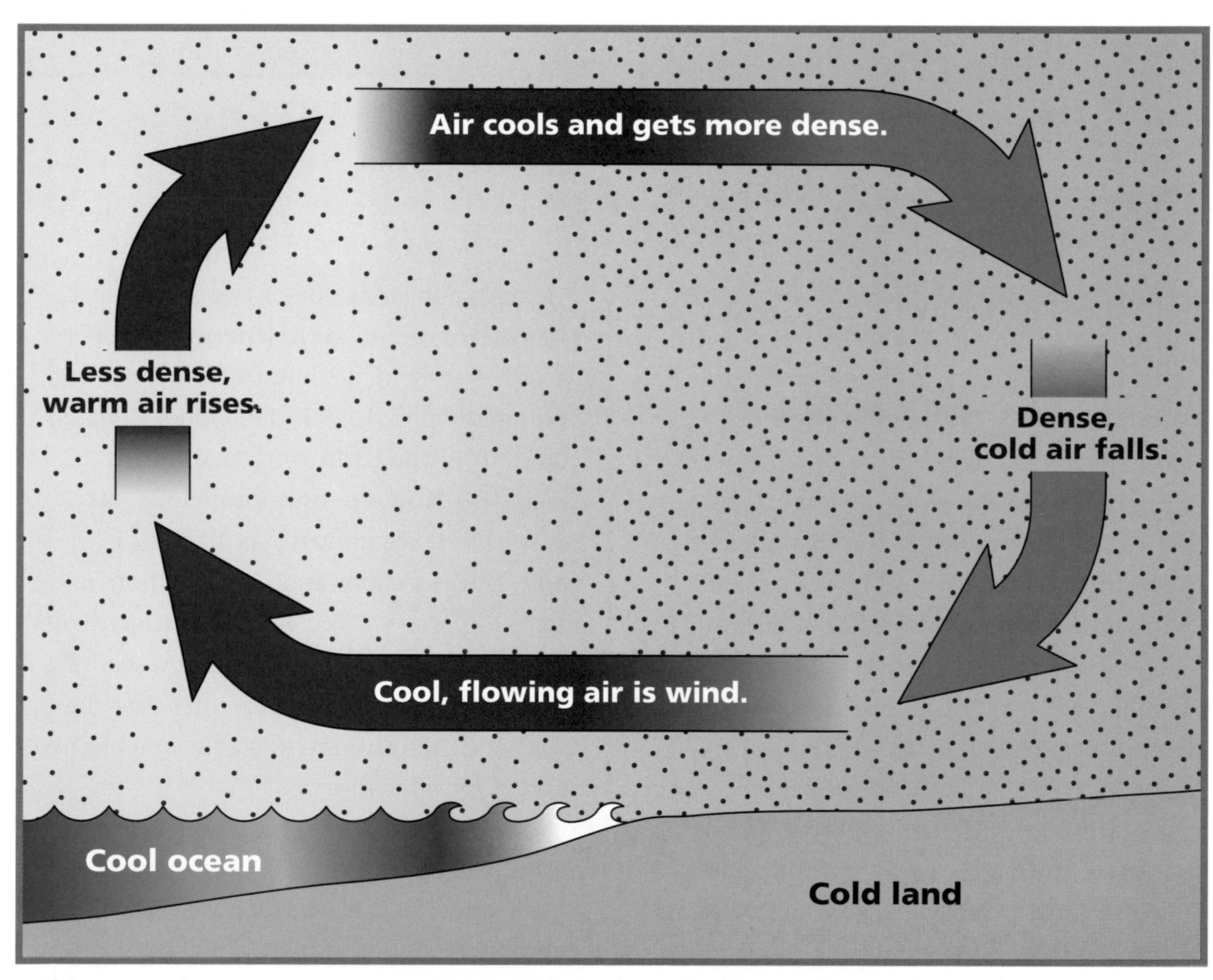

This convection current creates wind that blows from the land to the ocean.

Observing Clouds

Fluffy, puffy white clouds in a bright blue sky are one of the first memories of clouds that many of us have. You might remember a peaceful time lying on your back looking at the sky above, imagining shapes coming to life in the clouds. Ducks, people, trucks, houses, and horses might have paraded by as the puffs of bright cloud slowly changed.

You've already learned that clouds form when water vapor condenses on tiny surfaces of smoke, salt, and other **condensation nuclei**. But why do some clouds appear puffy and white and others grow to towering mountains? And what about those clouds that cover the sky as a gray, gloomy mass? Why are some clouds close to the ground and others faint streaks high above?

Clouds appear as one of two basic types, **cumuliform** and **stratiform**. *Cumuliform* describes the puffy, sometimes fast-moving and rapidly growing kind of cloud. *Cumulus* comes from the Latin word that means "heap." To grow a cumuliform cloud, air must be moving upward. As air rises, it cools. If water vapor and condensation nuclei are present, you've got the ingredients for a cumuliform cloud. When you see cumuliform clouds, you can infer that the weather conditions are unstable, and change may be in the works.

Stratiform clouds are flat and layered. *Stratus* is a Latin word meaning "layer." Stratiform clouds form when weather conditions are fairly stable. They result from the lifting of a large, moist air mass.

Meteorologists also observe where in the troposphere clouds form. High-level clouds form above 5,000 meters (m) and are given the *cirro-* prefix. Middle-level clouds form between 2,000 and 5,000 m and are given the *alto-* prefix. Low-level clouds form below 2,000 m. There is no special prefix for low-level clouds.

Some clouds may extend from low to high levels. They are nimbus clouds. *Nimbus* means "rain-bearing."

You can describe just about any cloud you observe by its shape and altitude. Here are some examples.

- An altostratus cloud forms a layer at a middle level.
- A cirrocumulus cloud is a puffy cloud that forms at a high level.
- A cumulonimbus cloud is piled up from low to high levels and brings rain.

Some low-level clouds have no prefix and are just known as stratus or cumulus clouds.

The words that describe clouds are very useful when you're recording weather observations and want to tell someone else what you have observed. Knowing why different clouds form gives you a good idea of weather conditions in your area.

Low-Level Clouds

Stratus

The base of stratus clouds is often around 600 m. Stratus clouds form in stable air. They appear flat and layered, with no lumps or bumps.

Luke Howard: Naming the Clouds

Luke Howard (1772–1864) is sometimes called the godfather of the clouds. Howard never trained as a scientist, but he loved nature, especially weather, from an early age. For more than 30 years, he kept a record of his weather observations. He presented his first system for classifying clouds in 1802. It is the same system meteorologists all over the world use today.

Howard also discovered that the air over cities is warmer at night than air over the countryside. Today we call this an urban heat island.

Stratocumulus

Stratocumulus clouds form when warm, moist air mixes with drier, cooler air. When this mixture moves beneath warmer, lighter air, it starts to form rolls or waves. It looks thick. It may bring drizzle or light precipitation.

Cumulus

Puffy white clouds at low levels are called cumulus clouds. When they are small and scattered, it means good weather. These are sometimes called fair-weather cumulus.

Cumulonimbus

Cumulonimbus clouds form on hot summer days. The sky may start clear, with little wind. Air heated by the ground rises. Convection cells form. Warm air rises through the cell center. Cooler air sinks down the sides. A cumulonimbus cloud forms. It has a base between 300 and 1,500 m. Rain starts to fall. Thunderstorms may develop.

Medium-Level Clouds

Nimbostratus

Nimbostratus clouds are sheet clouds carrying rain. Rain or snow falls almost continuously.

Altostratus

Altostratus clouds appear white or slightly blue. They can form a continuous sheet or look fibrous. They form between 2,000 and 5,000 m. Rain or snow may fall. Sometimes you can see the Sun through an altostratus cloud.

Altocumulus

Altocumulus clouds form between 2,500 and 5,500 m. They look like small, loose cotton balls floating across the sky.

Altocumulus mammatus

Altocumulus mammatus clouds look threatening, but actually indicate that the rainy weather is almost over. The clouds droop because the air is cooling and sinking.

Altocumulus lenticularis

Altocumulus lenticularis clouds are lens-shaped. Sometimes they look like flying saucers. They form at the top of a wave of air flowing over a mountain peak or ridge.

High-Level Clouds

Cirrus

Cirrus clouds are made of falling ice crystals. The wind blows them into fine strands. The longer the strands, the stronger the wind. Cirrus clouds indicate that the air is dry. Good weather should continue.

Cirrocumulus

Cirrocumulus clouds are high-level clouds composed of many smaller clouds. Some people think these clouds look like fish scales. A sky full of cirrocumulus clouds is sometimes known as a mackerel sky. It might mean that unsettled weather is on its way.

Cirrostratus

Cirrostratus clouds are high-level clouds that cover the sky. The cloud layer is thin and transparent. You can see the Sun or the Moon through cirrostratus clouds.

Jet Contrail

This jet is flying on a clear day. The contrail is formed by condensation of water vapor from the jet's exhaust. This could be called a humanmade cloud.

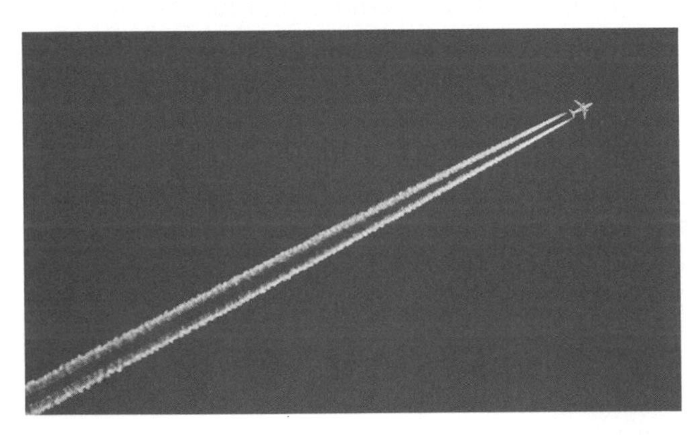

Think Questions

1. Look at the weather observations your class has recorded on the class weather chart.
   - If you included cloud observations, what were the most common types of clouds?
   - Try to identify any relationships between the types of clouds and other weather observations. For example, when air pressure decreased, did a certain kind of cloud appear?
2. If stratus clouds fill the sky for several days, what does that tell you about the stability of the air? What kind of weather might you expect?
3. Cumulonimbus clouds often form in the afternoon. What weather and land conditions might contribute to their forming later in the day? (Hint: Think about solar heating of Earth and heat transfer.)
4. Write and illustrate a short poem about clouds.

Weather Balloons and Upper-Air Soundings

In the late 19th century, meteorologists used kites to gather data about the air above them. Kites could fly up to 3 kilometers (km) into the air. Instruments on the kite gathered temperature, air pressure, and humidity data. Kites worked well when the wind cooperated.

The **radiosonde** was developed in 1943. A radiosonde is a weather-instrument package that can be carried into the stratosphere by a balloon. It has sensors for measuring temperature, relative humidity, and air pressure. Measurements are taken continuously as the balloon rises. A radio transmitter sends the data to a ground receiver. A tracking device monitors the location of the radiosonde during its flight. Wind speed and direction are calculated from the tracking data.

A weather balloon is made of a thin membrane of natural or synthetic rubber. It is inflated with either hydrogen or helium. A biodegradable plastic parachute is attached to the radiosonde. The balloon expands as it rises. When the balloon bursts, the radiosonde is carried to Earth by the parachute.

A radiosonde can be used as many as seven times. About one-third of the radiosondes launched by the National Weather Service (NWS) are recovered. Instructions are printed on each radiosonde, explaining how to return the device to the NWS. It goes to the NWS Instrument Reconditioning Branch in Kansas City, Missouri, where it is made ready for another flight.

In 1917, this box kite was used to measure humidity.

Preparing a kite to launch

There are more than 900 upper-air observation stations around the world, 108 of them in the United States. Most stations are located in the Northern Hemisphere. Observations from these stations are called soundings. Soundings are taken at the same times each day, 00:00 and 12:00 UTC (Universal Time Coordinated), every day of the year. The data are used for global and regional weather predictions, severe-storm forecasts, general aviation and maritime navigation, ground truth for satellite data, weather research, and climate-change studies.

New technology using the Global Positioning System (GPS) allows balloon-launchers to track down their device after landing and collect the equipment for reuse. However, radiosonde technology is still inexpensive and efficient enough to be used for the coordinated worldwide launches.

Think Questions

1. What is a radiosonde?
2. When do meteorologists launch weather balloons?
3. What are some of the advantages of using balloons to gather weather information?
4. Why do weather balloons expand as they rise through the troposphere?

Modern weather balloons are launched twice daily around the world. This balloon was launched in France.

Animal Rains

Can you imagine fish falling from the sky? How about spiders or frogs? It may sound like something in a science fiction novel or a fake news story, but reports of animals falling from the sky have been well-documented throughout history. In 1873, *Scientific American* reported that Kansas City, Missouri, was blanketed with frogs that dropped from the sky during a storm. In 1861, fish reportedly rained from the sky above Singapore. In 1901, residents described a rainstorm in Minneapolis, Minnesota, that produced frogs to a depth of several inches.

In 1948, golfers in England found their golf course covered in herring (a kind of fish) after a light shower. In 1953, Leicester, Massachusetts, was hit with a downpour of frogs and toads of all sorts, which

residents claimed choked rain gutters. In 1981, citizens in southern Greece awoke to small green frogs, native to North Africa, falling from the sky. And more recently, in 2007, hikers described watching spiders rain from the sky in Salta Province, Argentina.

Rain is actually water falling from the sky after evaporation and condensation. However, it is possible that these animals fell with the rain. Assuming that at least some of these accounts are reliable, what causes these bizarre phenomena?

Look back in your notebook. Can you think of any weather conditions you've studied in class that might help explain animals falling in rainstorms?

You might have guessed that wind played a part in this, and if so, you were right! The most accepted explanations involve strong winds traveling over water. These winds could pick up small creatures such as fish or frogs and carry them for up to several kilometers. Remember the videos about tornadoes and hurricanes that you watched at the beginning of the course? In those videos you learned that sometimes wind grows stronger and stronger in a cycle, causing surprising effects. That important concept is a key factor here.

Did you also guess that water was involved? You might have noticed that the animals listed typically live in or near water. One particular wind condition that forms over water is the most likely suspect for these animal rains.

Rare tornado-like waterspouts can be created when winds swirl into a strong vortex across a body of water. The vortex at the center of these storms could be strong enough to "suck up" surrounding air, water, and small objects like a vacuum. The vortex can carry objects in the air until cooling air forms clouds, and then precipitating water may drop the traveling objects.

Actual tornadoes have rarely been witnessed near animal rain events. Waterspouts might look less dramatic but potentially have winds that are just as strong, strong enough to lift up small objects and water. People have witnessed the falling animals, but there is no documented case of someone having witnessed the updraft part. This leaves quite a few questions about how the updrafts actually occur, and how high they go.

A waterspout on the ocean

The evidence, that most animal rains include small, light, aquatic animals, and that most accounts are preceded by a storm, support the updraft/waterspout hypothesis. However, these explanations do not account for why some animal rains lack water, why some accounts involve heavy fish or land animals, and why almost every account involves only one species, instead of a variety of similar-sized animals from a particular place. It is also remarkable that some reports say that the animals survive their plummets from the sky!

One eye witness from Texas reports, "My experience was coming out to my car after a short rain shower and I found small, partially frozen fish all over my car and the surrounding parking lot. The school where I worked was a couple of kilometers from a large lake. It is believed that a waterspout sucked up a school of the small herring-like fish that feed near the surface. The waterspout had carried them high enough in the atmosphere to partially freeze them. As the waterspout moved over land, it dissipated and the fish fell to the ground."

It is important to consider that witnesses could be biased or confused about what they saw. Some skeptics believe that if the rain is heavy enough, frogs that look like they're falling from the sky might just be jumping around in the rainstorm. In the middle of an intense storm, it could be hard to tell the difference. Indeed, other perceived animal rains have sometimes been explained later in other ways, including mass migrations of spiders (who can use their silk as parachutes to travel distances) and emergence of thousands of worms or frogs in unexpected areas.

But ultimately, any of these alternative explanations do not account for fish found far from any body of water, or the frozen frogs that hailed on Dubuque, Iowa, in 1882. A weather phenomenon is the likely explanation.

Will you see an animal rain in your lifetime? Now with so many people carrying around cell-phone cameras, maybe you will have the opportunity to see an animal rainstorm caught on video. Maybe you will be lucky enough to capture the footage yourself!

A frozen fish on the ground

Earth: The Water Planet

Earth is known as the water planet. Of all the planets in the solar system, Earth is the only one that has a vast ocean of water. If your first view of Earth from space was of the Pacific Ocean, you might think Earth was completely covered with water.

Earth is a closed system (almost). That means almost no material comes to Earth from space, and, on the flip side, no material is lost. This includes the water. Virtually all the water that is here now has been here for billions of years. The good news is that Earth's water is going to stay here for the next several billion years, but the probability of getting any more water from some outside source is essentially zero. What you see is what we've got.

Where Is Earth's Water?

By now, you know water is almost everywhere. It is in the ocean, in and on the land, and in the atmosphere. The pie charts show how water is distributed on Earth. A quick glance shows that just about all Earth's water is in the ocean. All that water and not a drop to drink, because to humans, sea water is toxic. To acquire the water we need for survival and to support civilization, we need access to the small portion of Earth's water that is fresh. After subtracting the amount of fresh water frozen as ice caps and glaciers and as groundwater, that small portion is less than 1 percent of the total water. Even this small percentage is still a pretty impressive volume of water, about 13 million cubic kilometers (km^3). This water is in lakes, rivers, swamps, soil, snow, clouds, water vapor, and organisms. It is known as free water because it is free-moving and constantly being refreshed and recycled on and over the land.

Most of the water we can easily use comes from rivers and lakes. Water in rivers and lakes is known as surface water. Water that falls as precipitation can either remain as surface water or seep into the ground, where it is stored in soil or porous rock. Underground water is known as groundwater. You can see from the pie charts that there is much more water stored underground than at the surface. It's water that is close at hand, but water that we can't see.

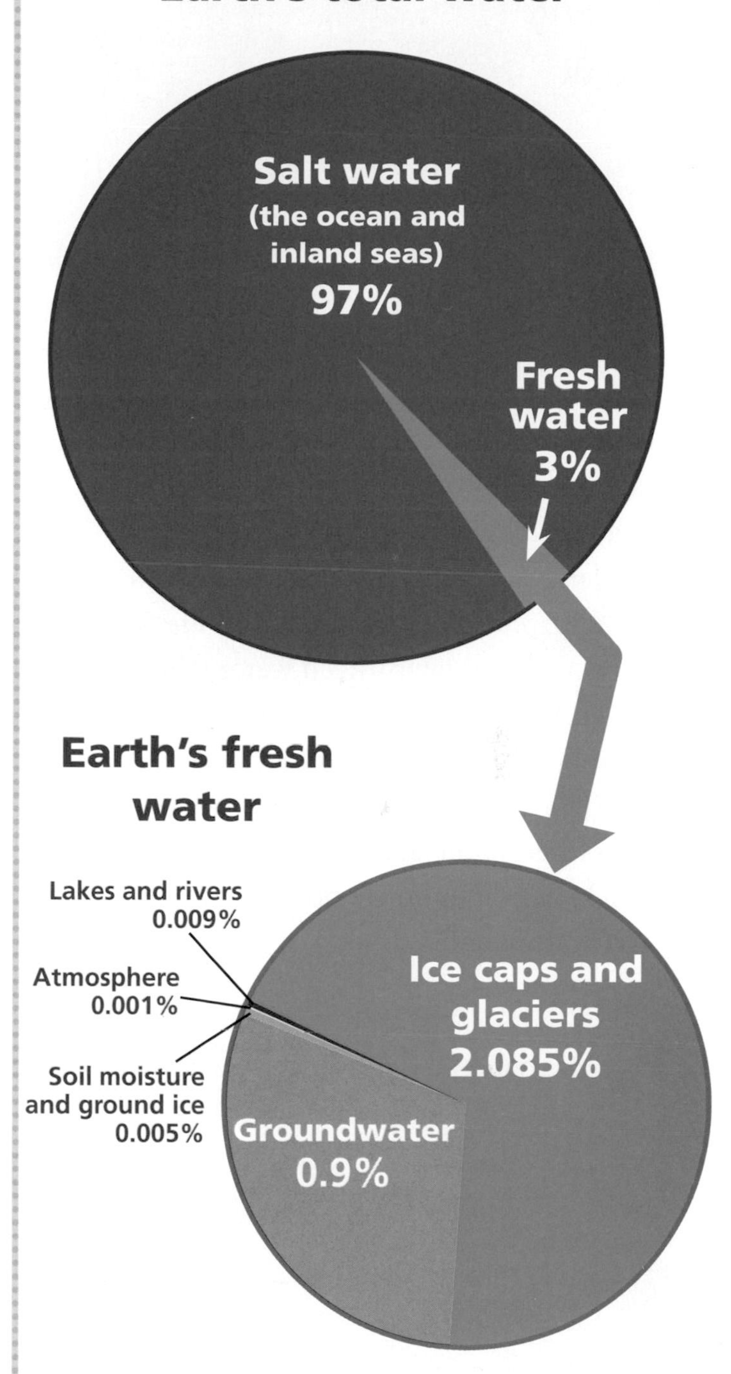

Water Use

Americans place high demands on water sources. Think about this. In 2005, the United States used a total of about 1,550 billion liters (L) of water a day. Over the course of a year, that adds up to more than 500,000 billion L! That translates into half a million km^3 per year. This is a significant percentage of the 13 million km^3 of free water available on Earth, about 4 percent.

People use water in many different ways. Most important, water is essential for life. Without water to drink, we wouldn't survive. And you can probably think of many nonessential ways you use water at home. You wash clothes, brush your teeth, and cook food with water. Swimming pools are filled with water, and lawns need water to grow. Humans also use water for transportation, for creating electricity, in manufacturing, and for agriculture. Many of these activities require good water quality. And, unfortunately, many of these activities create pollutants that can lower water quality.

A flooded cornfield in the Midwest

Water Distribution

Weather distributes water on Earth's landmasses. If weather did not continually resupply the land with water in the form of rain and snow, all land would be arid and lifeless. Weather does not, however, distribute water equally around the planet. Some places, like the Midwest, have variable water supplies, getting very little precipitation one year and a lot the next. During droughts, there may be severe water shortages, followed by floods. The deserts of the world are always parched, while the tropical rain forests are continually soaked. Adding to the unevenness of water distribution is the pattern of human-population distribution. Some densely populated areas, like Los Angeles, California; Phoenix, Arizona; and New York, New York, need more water than is available locally. They have to import water from faraway places.

Scientists are concerned about the warming trend on Earth. **Global warming** could affect both evaporation and precipitation in the United States. If more evaporation happens than precipitation, the land will dry, lake levels will drop, and river flows will decrease. Other regions may receive more precipitation than usual, creating floods and affecting vegetation. It will take worldwide planning and cooperation to adjust to the effects of global warming.

Earth is the water planet. Fortunately, water is one of our renewable resources. It is constantly being recycled among the atmosphere, land, and ocean. Humans can't change how much water there is. But we can make smart decisions about how much of it we remove from natural systems, how it is distributed, how it is used, and what happens to it after we use it. As the demand for water increases worldwide due to population increase and a rising standard of living (which requires water), everyone will have to participate in water conservation. Industries will need to be more efficient with water use and careful not to introduce pollutants into water sources. Agriculture will have to develop more conservative crop-watering practices. And every citizen will have to become more aware of the value of water and treat it as the most precious substance on Earth.

For more information about Earth's water, visit FOSSweb.com.

Think Questions

1. Earth's human population reached 7 billion in 2012. It is estimated that it will reach between 8.3 and 10.9 billion by 2050. What impact might that increase have on freshwater sources?
2. If humans use more fresh water from lakes and groundwater sources, how might that affect the rest of the water cycle? Earth's systems? Other organisms living on Earth?
3. As Earth's human population continues to grow, what are some strategies that might help cut back on the amount of fresh water used by humans?

Ocean Currents and Gyres

In previous investigations, you learned what happens when the Sun's radiant energy heats Earth's surface. The ocean covers a whopping 70 percent of Earth's surface. As the ocean heats or cools, how will all that water move? And how will those movements affect weather and climate?

The Driving Forces of Ocean Currents

Just as there are global wind patterns (as we explored in Investigation 6), there are global water patterns, called **ocean currents**. There are three main forces that affect ocean currents: winds, differences in water density, and tides.

Winds. Winds drive currents near the ocean's surface (approximately the upper 20 percent of the ocean). In coastal areas, local winds can drive currents, but in the open ocean, global winds drive currents.

Differences in water density. Deep ocean currents (the lower 80 percent) are driven mostly by differences in density of water. You learned earlier that convection currents form when fluids have different densities. What are two factors that could cause ocean water to have different densities? *Hint: You experimented with each factor in Investigation 4.*

Which heats up more quickly, water or land? As you learned in Investigation 5, water heats and cools slowly, but even these slow changes will cause temperature differences in the ocean from one area to another. As the Sun's radiant energy warms the ocean, differences in temperature will cause differences in density, resulting in convection currents.

Which is more dense, a solution with lots of salt or little salt? As you learned earlier, solutions with more salt are more dense. Variations in the concentration of salt, called **salinity**, in ocean water cause differences in density, resulting in currents.

The combined effect of density differences due to temperature and salinity result in upward and downward flow of water masses. These currents occur at both deep and shallow ocean levels and move much slower than surface currents.

The effect of temperature and density on currents is greatest in Earth's polar regions. As polar sea ice forms, the surrounding seawater gets saltier. This is because pure water freezes and separates from the salt, leaving all of that dissolved salt in the ocean. Water near the polar regions sinks from being more dense not only from cold, but also from being saltier. As this cold, salty water sinks, the surrounding surface water moves in to replace it. Once the water sinks, it travels through the ocean basins, eventually coming back to the surface, about 1,500 years later! This motion drives the deep-ocean currents that weave their way around the planet in a system we call the Great Ocean Conveyer Belt, a huge convection current.

Tides. The third main force that affects ocean currents is tides. Tides are driven by the gravitational force of the Moon and Sun acting on the rotating Earth. Near shore, the rise and fall of water interacts with the shapes of shorelines to create surface tidal currents. Tidal currents follow regular patterns: flooding when the tide moves toward the land and ebbing when it moves toward the sea.

The shape of the shoreline can dramatically affect tidal current. The Bay of Fundy between Nova Scotia and New Brunswick is a classic example, with tides over 15 meters (m), the highest in the world. As you might well imagine, aquatic ecosystems in coastal areas are dramatically affected by daily tidal currents.

Rocks on the shore in Nova Scotia at high tide and low tide

The movement of ocean water is a complicated thing. Factor in the tides, density changes due to differences in temperature and salinity, local winds and global winds, and the Coriolis effect (caused by rotation of Earth), and you have most of the pieces of the ocean current puzzle. What's the final puzzle piece? The ocean is broken up by giant landmasses of continents and islands. Just as water in a stream will flow differently if you drop in a large rock, the movement of ocean water is affected by the geology of Earth.

The flow of ocean water is affected by the landmasses it encounters.

Local Currents

The patterns of ocean currents affect how fish and other sea life move around the ocean, control how ships travel through the water, and sometimes carry swimmers far from the beach if they aren't careful. When surface currents run close to shore, a visitor to the beach could experience a **rip current**. These local currents form around low spots, breaks in sandbars, or structures such as piers. Rip currents can move faster than even Olympic swimmers can swim, and if you are caught in one, don't fight it. They are generally not more than 25 m wide, and they usually break up not far from shore, so it's best to swim around them.

Global Current Patterns

Ocean currents can form swirling areas called **gyres**. A gyre (rhymes with "tire") is a large system of rotating ocean currents driven largely by winds. There are five main gyres found in Earth's ocean, as well as other less-major gyres such as the one that circles the South Pole.

What causes these gigantic whirlpools in the ocean? Do you notice any patterns in the map of gyres? All of the major gyres occur in the middle of the spaces between continents. As ocean currents hit shallower areas of coast, they start to move in a different direction. Think of the water as sweeping along the coastline in a large-scale current, called a **boundary current**. As the water approaches a continent and then sweeps in a different direction, you get the first part of a swirl that will ultimately form a gyre.

A boundary current of the South Pacific Ocean is the Humboldt Current (or Peru Current) that moves northward along the coast of Chile. As it veers away from Peru, deeper water upwells to replace the

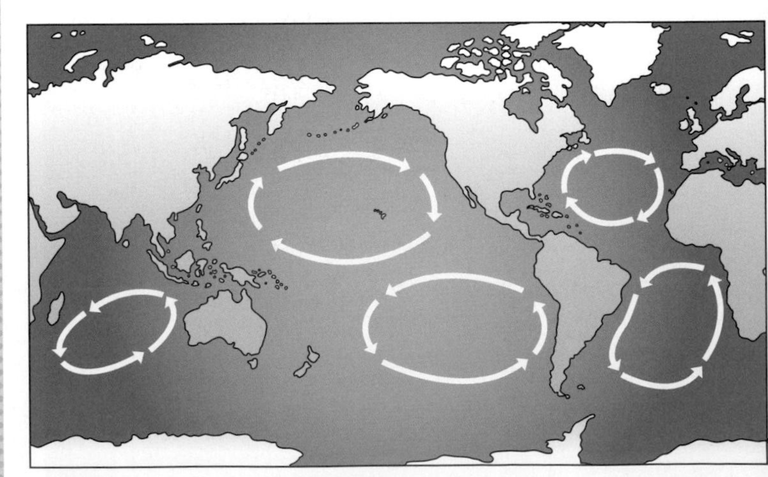

The five main gyres found in the ocean

westward-moving water. This is one of the biggest upwelling systems on Earth. It brings up cold, nutrient-rich ocean water, feeding an abundance of marine life and the largest fishery on Earth. About 20 percent of Earth's fish catch is made possible by the Humboldt Current! Because the current is cold, it cools the weather and climate of Chile and Peru.

Perhaps the most familiar of the boundary currents is the **Gulf Stream** in the North Atlantic. It goes northward along the east coast of the United States and then across the North Atlantic Ocean toward Europe. The warm water of the Gulf Stream affects not only the climate of the east coast of North America from Florida to Newfoundland, but also the west coast of Europe, making for much milder climates than would exist otherwise.

How do you think ocean currents might affect your local climate? Record your answers in your notebook.

The Great Pacific Garbage Patch

We see it every day. Someone carelessly drops a plastic cup or wrapper on the ground as they walk, or tosses it out their car window. Where are those plastic objects now? Plastic doesn't biodegrade (break down in nature) quickly, and trash on the ground can get washed by rainwater into storm drains that eventually lead to the ocean.

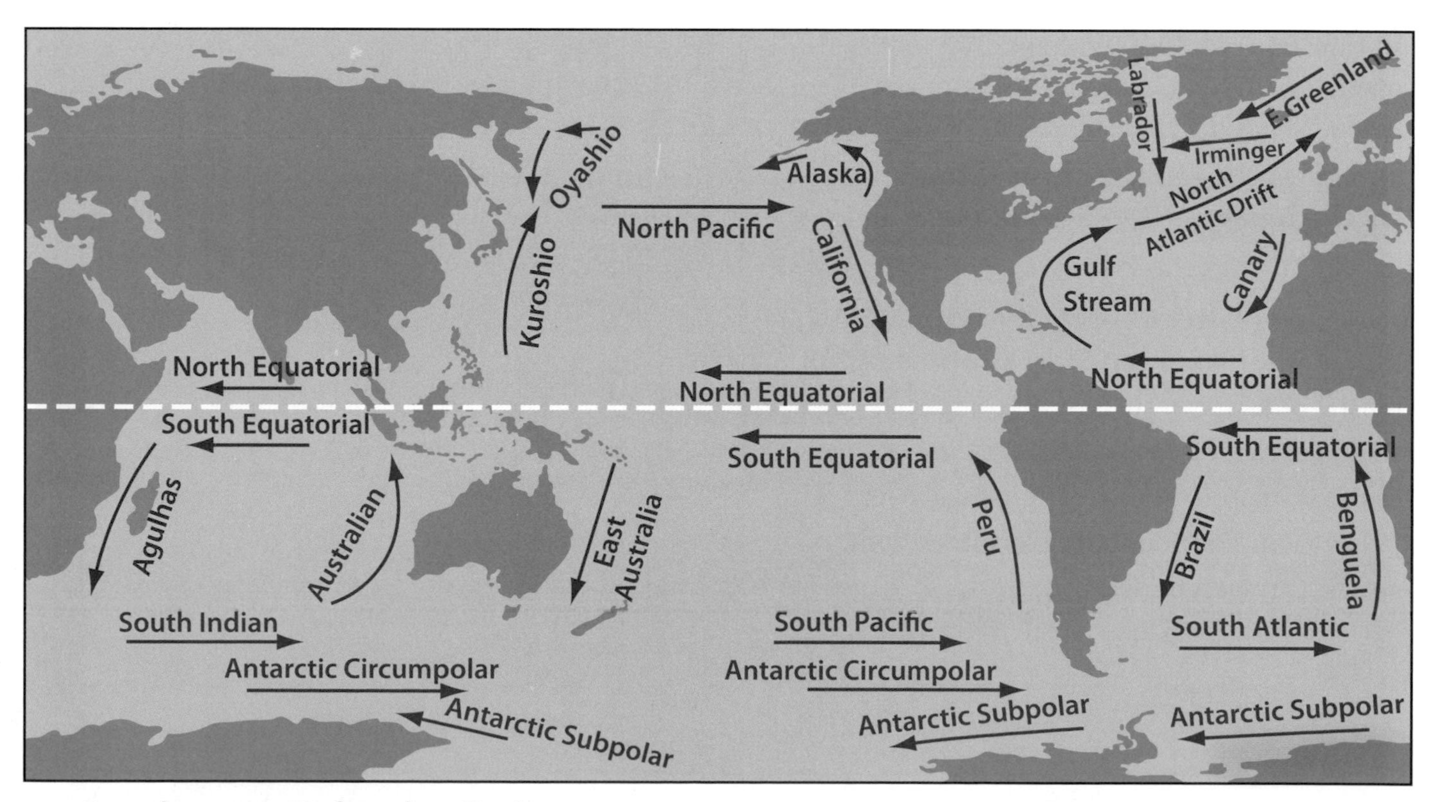

Boundary currents found on Earth

Imagine all that plastic trash floating into the ocean and getting transported far from its source. Where will it end up?

About 1,600 kilometers (km) northeast of Hawaii in the Pacific Ocean is the Great Pacific Garbage Patch, a huge swirling current of garbage, mostly plastic. The pieces can be as large as huge fishing nets but are mostly plastic pieces the size of corn flakes or rice grains, broken from larger plastic objects. This giant oceanic whirlpool of garbage doubles in size every decade. At the end of 2009, it was estimated to be roughly twice the size of Texas! The garbage is caught in what oceanographers call the North Pacific Ocean Gyre, one of the giant gyres generated by the world's ocean currents. The garbage includes toxic chemicals.

Plastic doesn't biodegrade quickly, but it will break down into tinier and tinier pieces, just the right size for marine life to eat. Fish that feed on plankton eat the tiny plastic particles, and toxic chemicals from the plastic leach into fish tissue. Birds such as albatross forage in the North Pacific and eat large amounts of plastic, including fishing line, light sticks, and lighters. Some birds can purge themselves of the plastic by throwing it up, but thousands die each year from eating plastic debris. Sea turtles, many species of which are endangered (at risk of becoming extinct), may think floating plastic bags are jellyfish and accidentally consume them, which can lead to death.

Think Questions

1. What are three forces that drive ocean currents?
2. What are the two main factors that affect deep ocean currents?
3. Does the map of major ocean currents look anything like what you would have predicted? How does it differ from what you would expect?
4. Look at the map of major boundary currents on page 81. What would be the best direction to start out for sailors making these trips?
   - Spain to Florida
   - China to California
   - Australia to Africa
5. Have you ever seen someone litter plastic? What is a possible solution to prevent people from littering?
6. What are possible health risks to humans due to plastic trash ending up in the ocean?

El Niño effects can cause drier than normal conditions.

El Niño

Ocean currents have a big effect on weather and climate. The interplay between ocean currents, weather, and climate can be quite dramatic.

Usually wind blows strongly from east to west along the equator in the Pacific, causing water to pile up about half a meter high in the western Pacific. Simultaneously in the eastern Pacific, cold deep water flows upward to replace the warmer surface water that has been blown to the west. So normally the western Pacific has surface water about 8 degrees Celsius (°C) warmer than the eastern Pacific surface water, which is about 22°C.

Every few years, this balance changes in an event called **El Niño**. El Niño starts when the equatorial winds get weaker and some of the warm water that has piled up in the western Pacific starts to slip back along the equator to the east. As a result, not as much cold water upwells in the eastern Pacific, and the ocean surface temperatures are warmer than usual in the eastern Pacific. These warm surface temperatures are one of the hallmarks of El Niño.

El Niño has dramatic effects on climates worldwide. There are heavier than normal rains in Peru and Ecuador. Less upwelling of cold, nutrient-rich water reduces fish populations. In the northern United States (1 on map below), winters are warmer and drier than average, with less snow. Cooler and wetter-than-average winters occur in northeast Mexico (2) and the southeast United States. East Africa (3) is wetter than normal from March to May, but south-central Africa (4) is drier than normal from December to February. Drier conditions occur in parts of Southeast Asia and northern Australia (5), increasing bush fires, worsening haze, and causing the dramatic decrease of air quality. El Niño can cause a wetter, cloudier winter in northern Europe and a milder, drier winter in the Mediterranean region.

Visit FOSSweb for more information about El Niño.

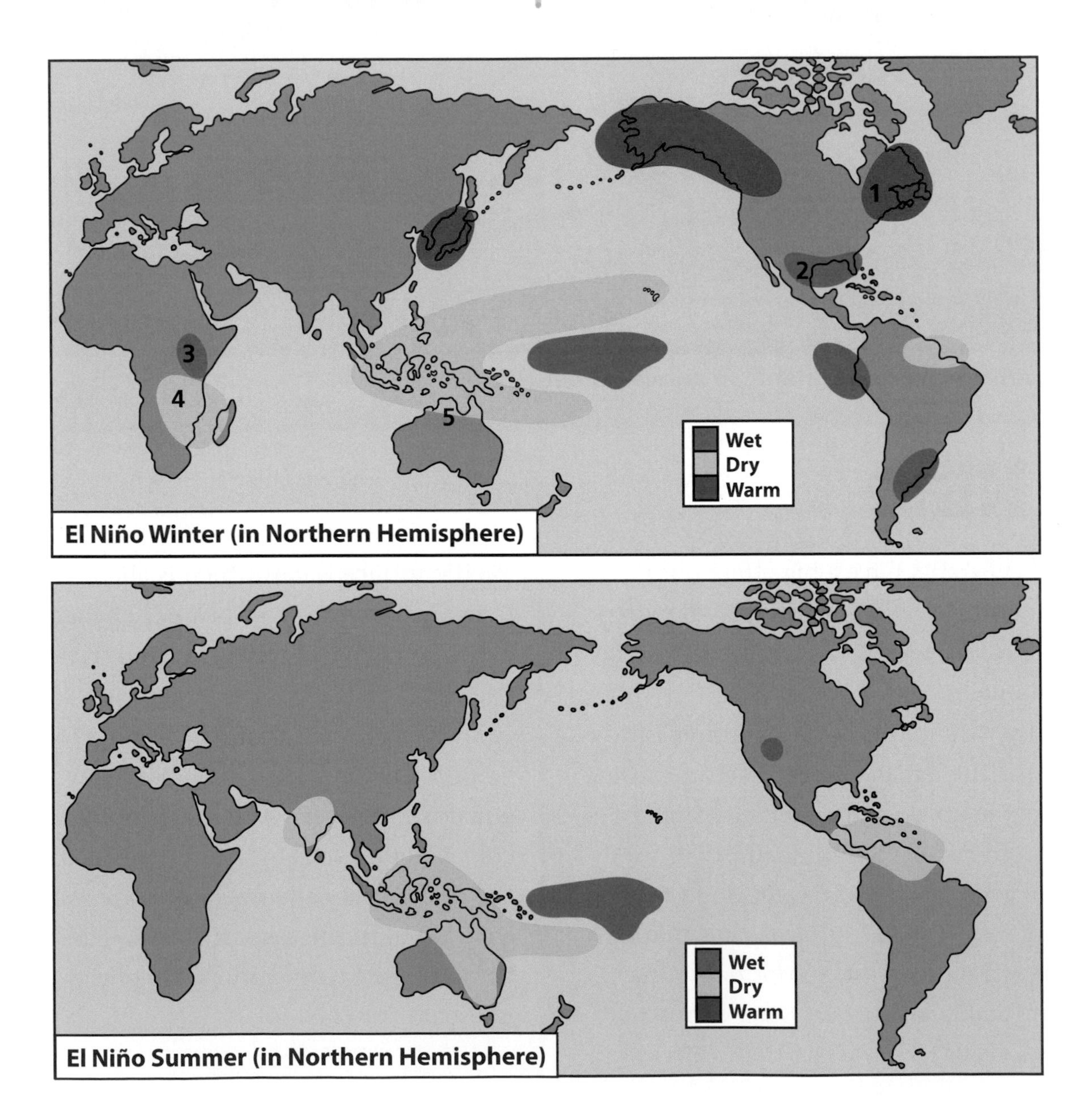

Climate change can cause glaciers to melt more quickly than they otherwise would.

Climates: Past, Present, and Future

Climate change is in the news and generates a lot of questions. Is it real? Is it natural or caused by humans? Is it happening fast? What will happen if we don't stop it? What can we do to stop it?

The answer to the first question is yes, climate change is real. The evidence is in temperature data over time. The average temperature of Earth's lower atmosphere (troposphere) has gone up by about 0.6 degrees Celsius (°C) over the past 100 years. This may not seem like much, but some weather effects result from even slight temperature changes. The temperature of the upper ocean layer [top 300 meters (m)] has increased by more than 0.3°C in the past 40 years. Even a small increase in global ocean temperature represents a lot of heat.

One of the major concerns about a temperature increase is that global warming can cause ice sheets and glaciers in or near polar regions to melt at an increasing rate. This extra liquid water added to the ocean can raise the level of the ocean. Farms, villages, towns, and cities built at or near sea level could be devastated by the rising water.

Retreat of South Cascade Glacier in Washington State

The answer to the second question (is it caused by humans?) is fairly straightforward also. Although natural factors can cause global warming or cooling, the most significant factor in the *current* warming trend is a human-caused increase in greenhouse gases that absorb infrared rays, trapping heat in the atmosphere. Two ways that we are increasing greenhouse gases are (1) burning fossil fuels such as coal, oil, natural gas, and gasoline that produce carbon dioxide (CO_2) as a product during combustion, and (2) cutting down forests, which through photosynthesis remove carbon dioxide from the atmosphere and hold it in the form of cellulose for a very long time.

Cars send a lot of carbon dioxide into the atmosphere.

Earth Systems

The answers to the other questions are a lot more complicated. Scientists are watching Earth systems closely to see how fast changes are happening. Here are some observations.

Ice is melting. Ice melting is a predicted result of global warming. Large chunks of ice are breaking off the Antarctic ice sheets. There is less ice at the North Pole than there was years ago. Many glaciers (large sheets of ice) are getting smaller. At the present rate of melting, glaciers in Montana's Glacier National Park will be gone in 50 years. This also affects permafrost, which is soil that stays frozen, so it is solid enough to build on. When permafrost begins to melt, there are big problems for people and other organisms who live on it.

Permafrost that has started to crumble along a shoreline

Sea level is rising. Over the last 100 years, sea level has risen 15–21 centimeters (cm). The increase in volume comes from melting ice and from heat, which makes water expand. When sea level rises, tides go farther up the beach, and the ocean permanently submerges coastal areas. A large percentage of humans live in large cities near coasts, with island dwellers in particular danger. Also, much of the world's fertile cropland is in low-lying areas.

When sea level rises, coastal areas are submerged in water.

Visit FOSSweb for a link to a website that lets you change sea level and witness the effects.

Ecosystems are changing. As ice and permafrost melt and the ocean warms slightly, nonhuman organisms in coastal or permafrost areas are affected as well. Some species may not have suitable habitats to live in anymore. Others may try to migrate and end up affecting other ecosystems nearby.

What's the difference between global warming and climate change? Global warming is simply an increase in the average temperature of Earth. Climates change as a result of that. Predicted changes include those mentioned above as well as droughts, increased frequency and intensity of storms, hurricanes, and flooding. Another effect is the change in the onset and duration of seasonal weather impacting plant and animal life cycles.

What's in the Future?

We can be fairly certain that Earth's temperature will continue to rise as long as the concentration of greenhouse gases in the atmosphere continues to rise. You saw the relationship between carbon dioxide and global temperature in the graph you studied in class. As we continue to add greenhouse gases like carbon dioxide to the atmosphere, we can expect a continuously warming global climate. We can affect that future significantly if we choose to stop adding greenhouse gases to the atmosphere

How can humans reduce their carbon dioxide emissions?

Carbon Dioxide Concentration for the last 800,000 Years
(in parts per million, measured from trapped bubbles of air in an Antarctic ice core)

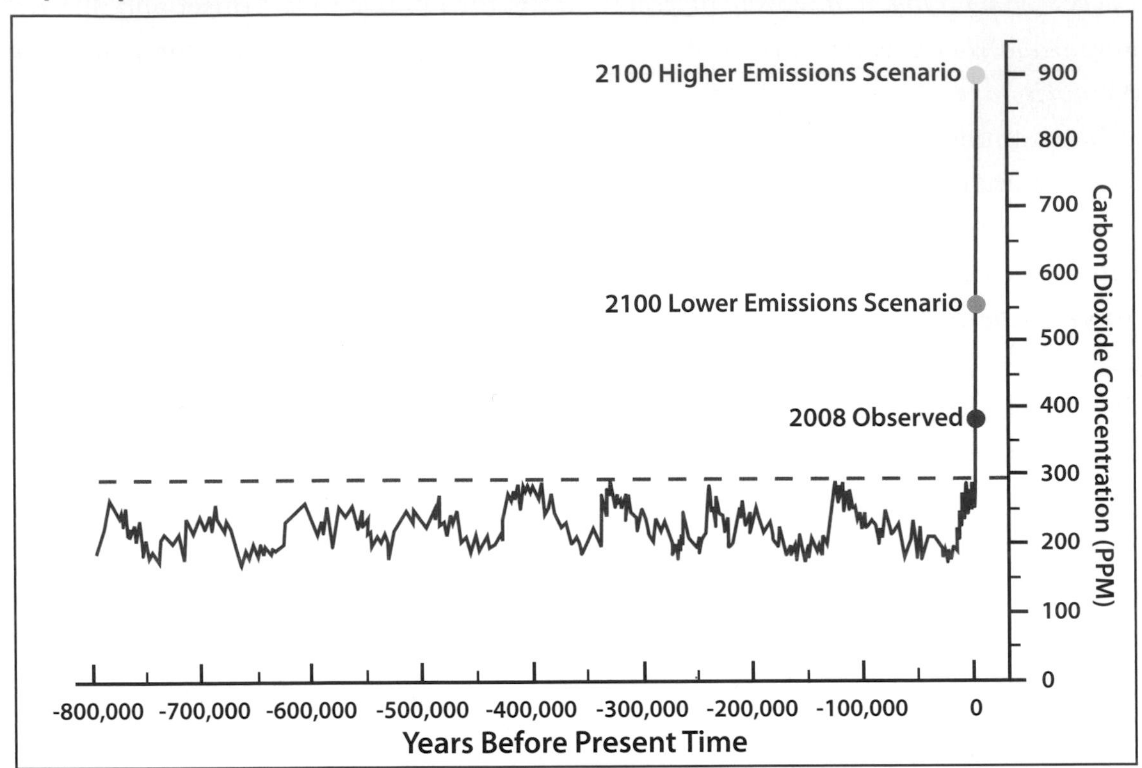

For more information about this data, visit FOSSweb.

and restore forests that have been burned or stripped away.

The graph above shows atmospheric carbon dioxide over time and projects into the future. Reductions in carbon dioxide emissions by humans will result in the "lower emissions scenario" by the year 2100. If there are no significant changes in human behavior, the "higher emissions scenario" shows the predicted 2100 levels. Will we be at the "lower emissions" mark or the "higher emissions" mark by 2100? That is up to us.

Change won't be easy, but it is possible. Individuals can help by conserving energy to reduce greenhouse-gas emissions. Carbon dioxide is released whenever humans start a gasoline engine or use electricity generated in power plants that burn coal, gas, and oil. That means you are very likely generating greenhouse gases whenever you use the air conditioner, turn on a light, play a video game, charge your cell phone, or travel by car. As individuals, we can conserve electricity, use public transportation, and produce some of our own food in a garden. As a society, we can develop new technologies that don't require fossil fuels and emit carbon dioxide (including harnessing solar, wind, and water-powered energy), build better public transportation systems, and make products that are far more efficient than those we have today.

Think Questions

1. What do you do that contributes to carbon dioxide gas emissions?
2. What could you do to reduce your carbon dioxide emissions?

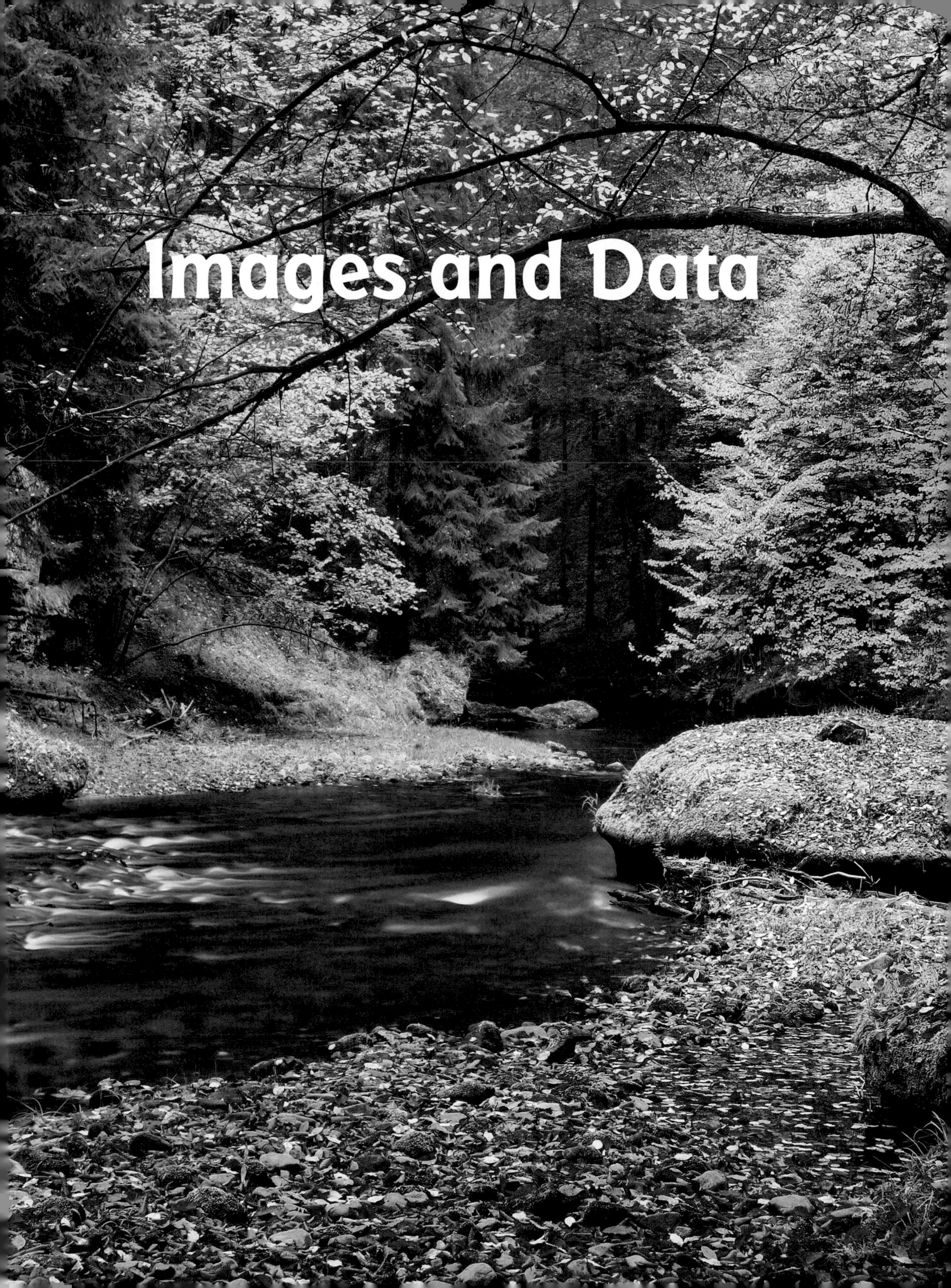
Images and Data

Images and Data Table of Contents

World Map

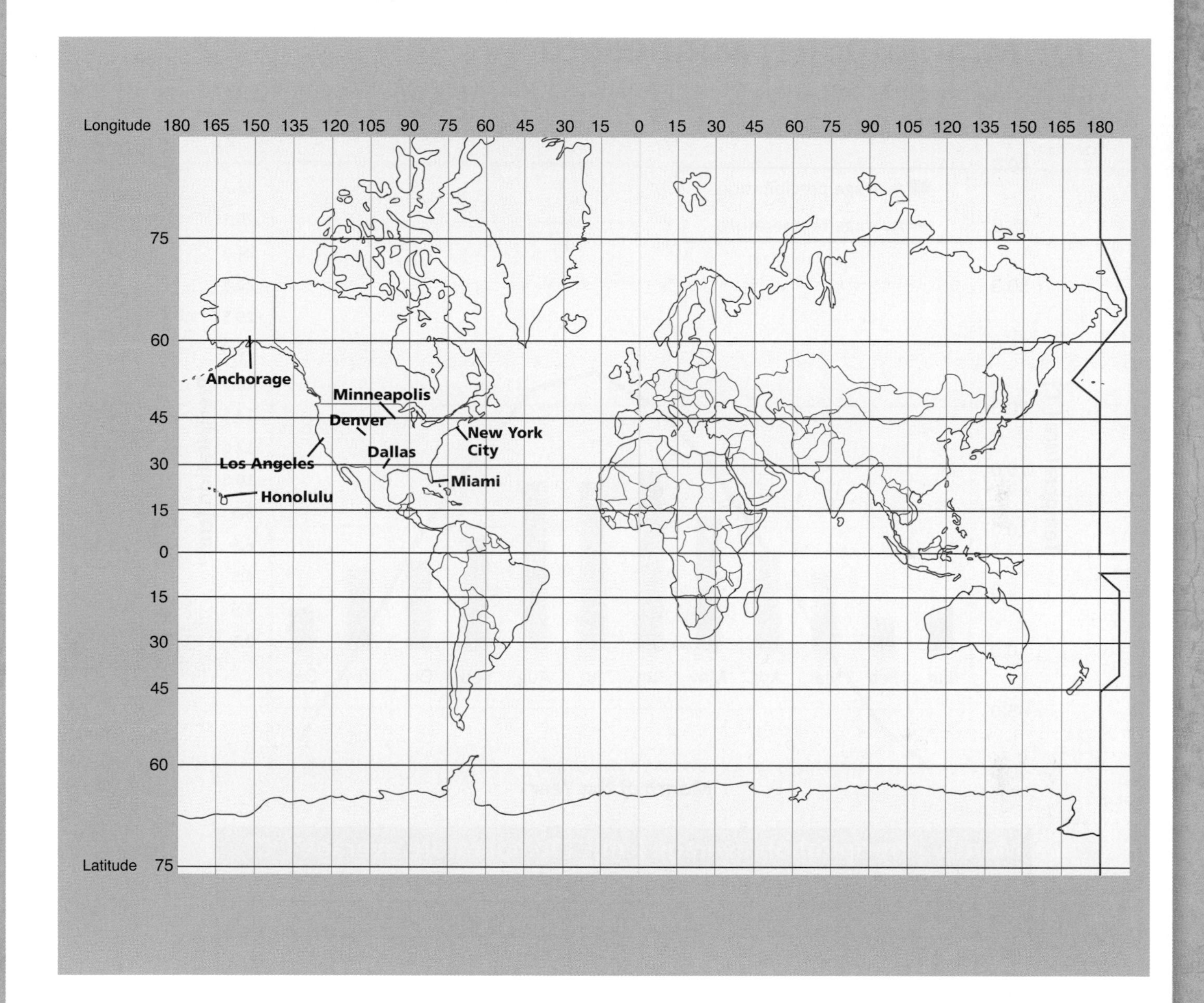

Longitude 180 165 150 135 120 105 90 75 60 45 30 15 0 15 30 45 60 75 90 105 120 135 150 165 180
75
60
45
30
15
0
15
30
45
60
Latitude 75
Anchorage
Minneapolis
Denver
New York City
Dallas
Los Angeles
Miami
Honolulu

Minneapolis-Area Climate

Climate Graph and Data for 30 Years for Minneapolis, Minnesota

Latitude: 45.0° N Longitude: 93.3° W

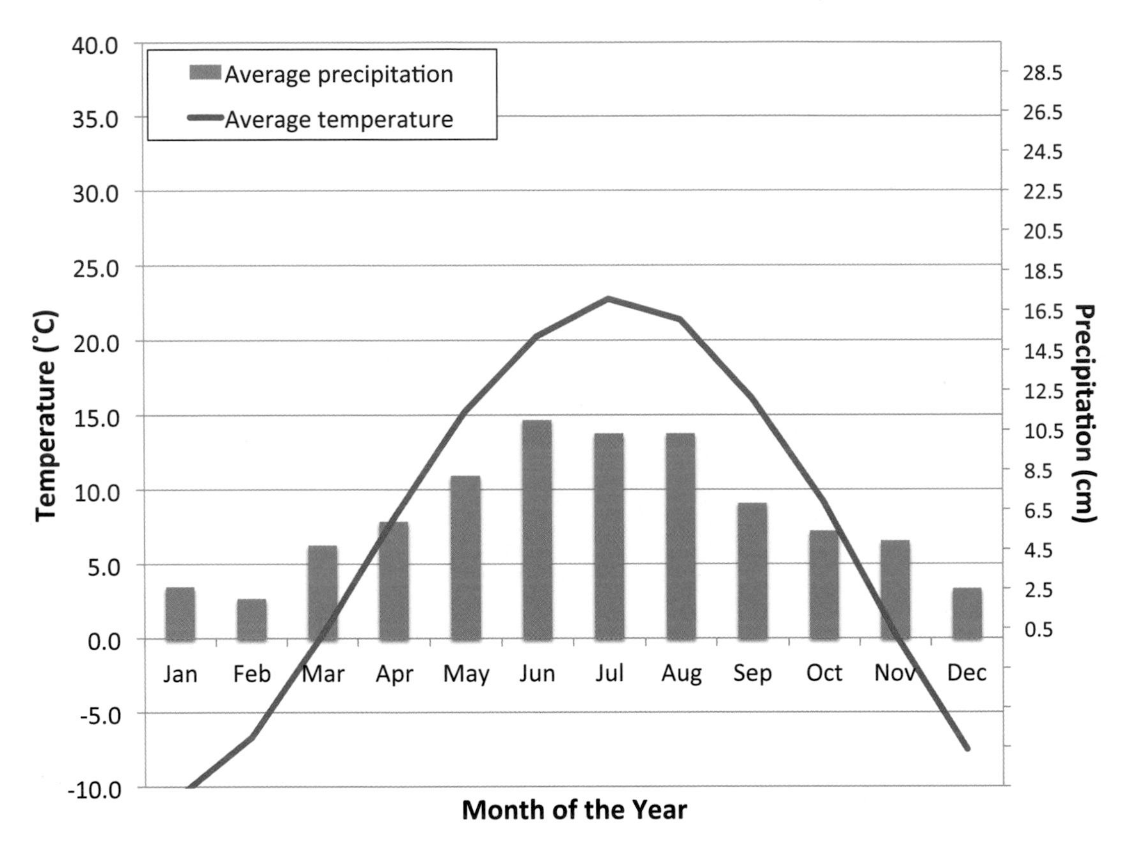

Month	Precipitation average (cm)	Temperature average (°C)
Jan	2.6	-10.6
Feb	2.0	-6.7
Mar	4.7	0.3
Apr	5.9	8.1
May	8.2	15.3
Jun	11.0	20.3
Jul	10.3	22.8
Aug	10.3	21.4
Sep	6.8	16.1
Oct	5.4	9.2
Nov	4.9	0.3
Dec	2.5	-7.5

Miami-Area Climate

Climate Graph and Data for 30 Years for Miami, Florida

Latitude: 25.8° N Longitude: 80.2° W

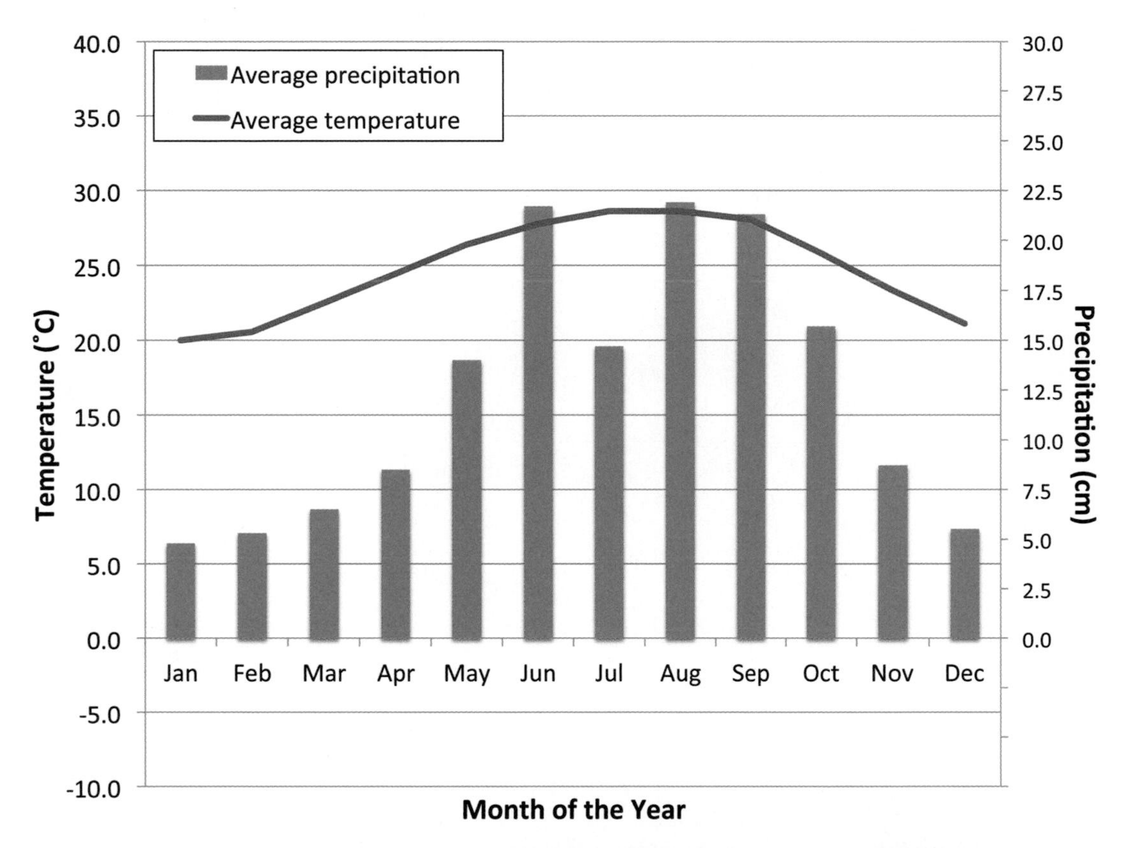

Month	Precipitation average (cm)	Temperature average (°C)
Jan	4.8	20.0
Feb	5.3	20.6
Mar	6.5	22.5
Apr	8.5	24.4
May	14.0	26.4
Jun	21.7	27.8
Jul	14.7	28.6
Aug	21.9	28.6
Sep	21.3	28.1
Oct	15.7	25.8
Nov	8.7	23.3
Dec	5.5	21.1

US Map with City Locations

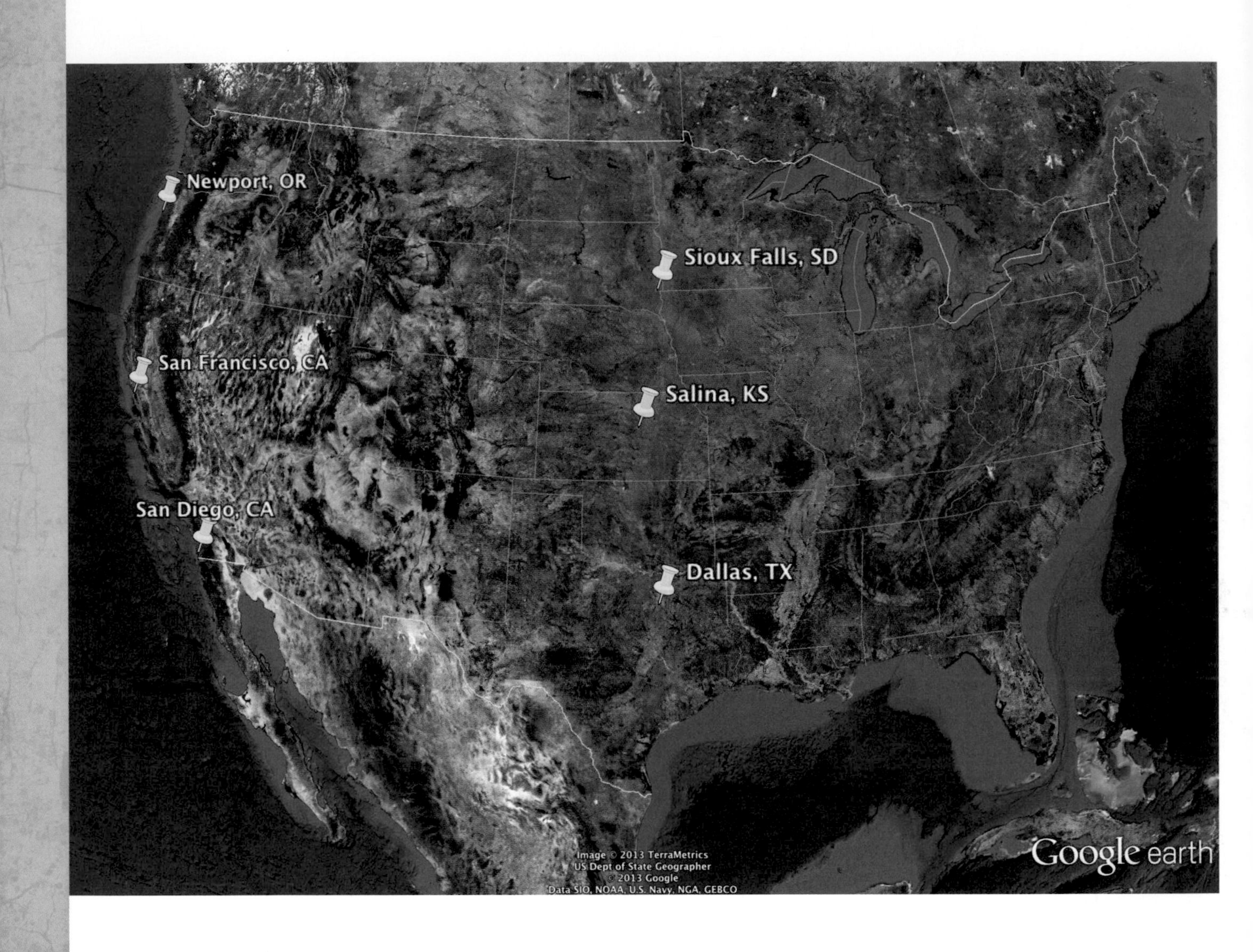

Great Plains Cities: Climate Data over 30 Years

Sioux Falls, South Dakota

Latitude 43.5° N
Longitude 96.7° W
Elevation 430 m

Average temperature in °C

Jan	Feb	Mar	Apr	May	Jun
-10.0	-6.1	0.3	7.8	14.4	20.0

Jul	Aug	Sep	Oct	Nov	Dec
22.8	21.4	16.1	8.9	2.5	-7.5

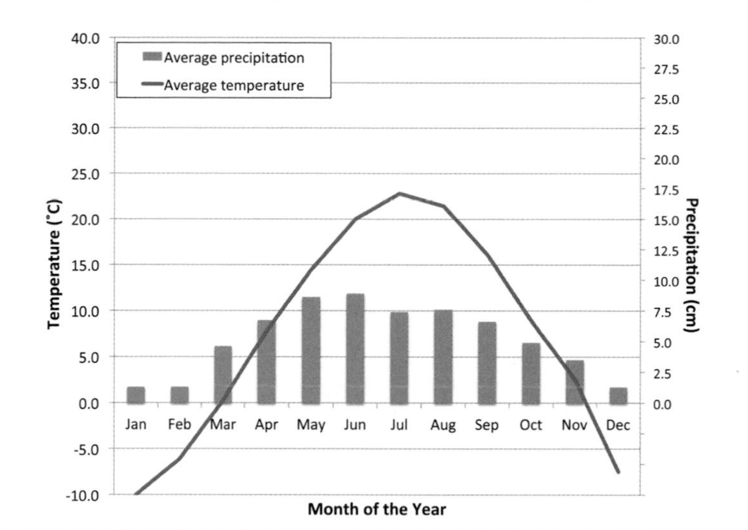

Salina, Kansas

Latitude 38.8° N
Longitude 97.6° W
Elevation 370 m

Average temperature in °C

Jan	Feb	Mar	Apr	May	Jun
-1.7	1.7	7.2	12.8	18.1	24.2

Jul	Aug	Sep	Oct	Nov	Dec
27.2	26.4	21.1	14.4	6.4	0.3

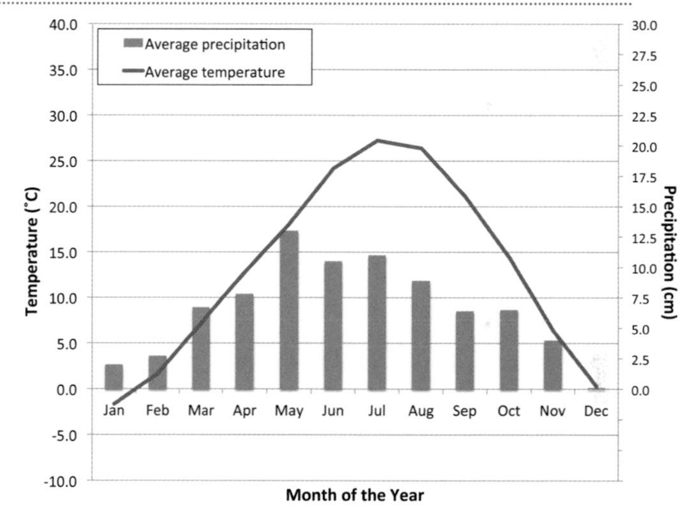

Dallas, Texas

Latitude 32.8° N
Longitude 96.8° W
Elevation 160 m

Average temperature in °C

Jan	Feb	Mar	Apr	May	Jun
7.5	10.6	15.0	18.9	23.6	28.1

Jul	Aug	Sep	Oct	Nov	Dec
30.3	30.0	26.1	20.3	13.6	8.9

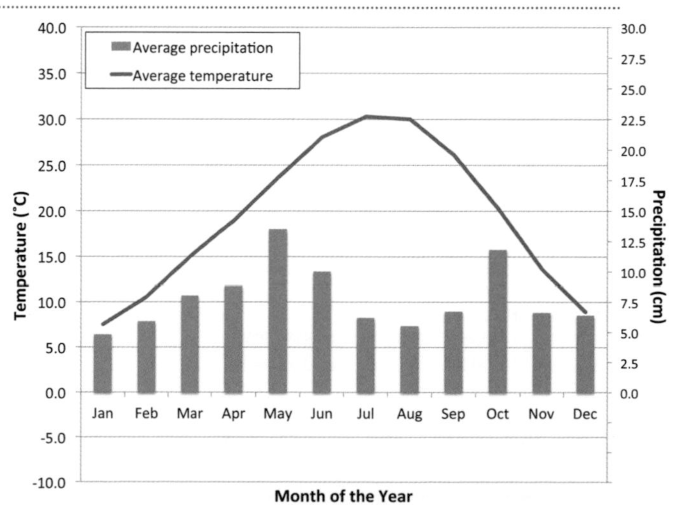

West Coast Cities: Climate Data over 30 Years

Newport, Oregon

Latitude 44.6° N
Longitude 124.1° W
Elevation 3 m

Average temperature in °C

Jan	Feb	Mar	Apr	May	Jun
7.2	7.8	8.6	9.4	11.4	13.1

Jul	Aug	Sep	Oct	Nov	Dec
14.4	14.7	13.9	11.7	9.2	7.2

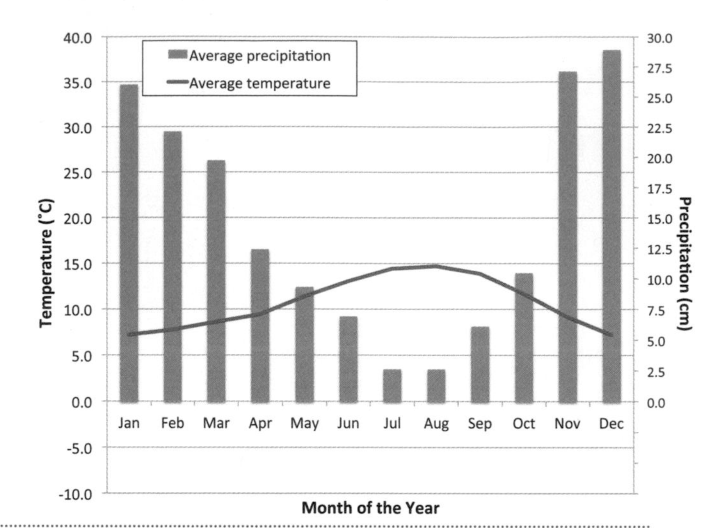

San Francisco, California

Latitude 37.8° N
Longitude 122.4° W
Elevation 3 m

Average temperature in °C

Jan	Feb	Mar	Apr	May	Jun
11.1	12.8	13.1	14.2	14.4	15.8

Jul	Aug	Sep	Oct	Nov	Dec
16.1	16.9	17.5	16.9	14.2	11.7

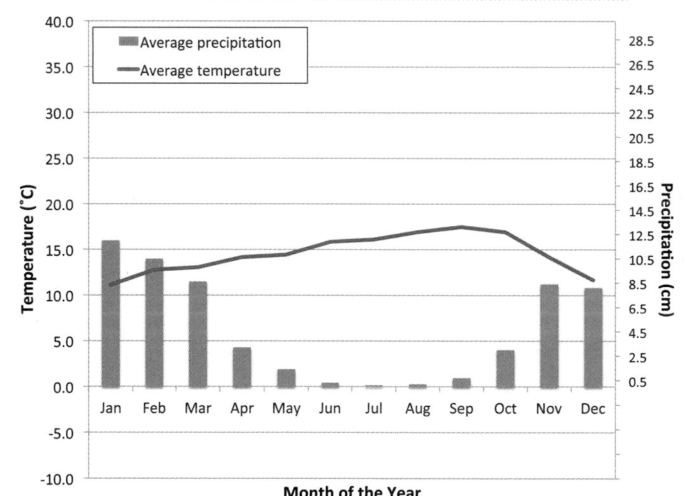

San Diego, California

Latitude 32.7° N
Longitude 117.2° W
Elevation 3 m

Average temperature in °C

Jan	Feb	Mar	Apr	May	Jun
13.1	13.3	13.9	15.3	16.9	19.2

Jul	Aug	Sep	Oct	Nov	Dec
21.7	22.2	21.7	18.9	15.6	13.1

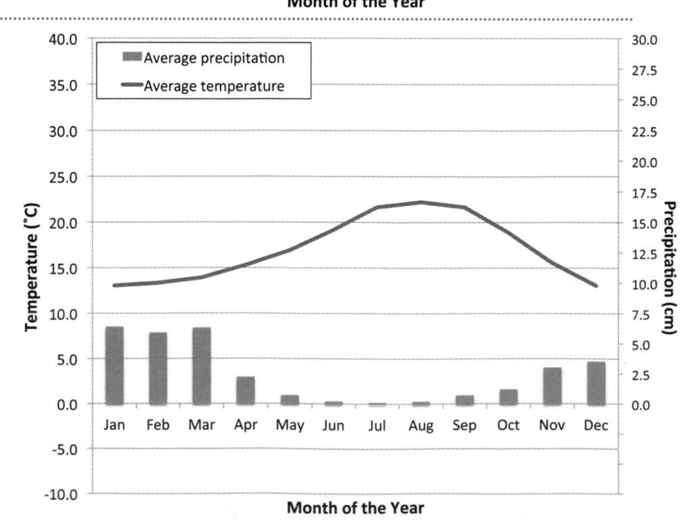

Radar Images of Cloud Cover

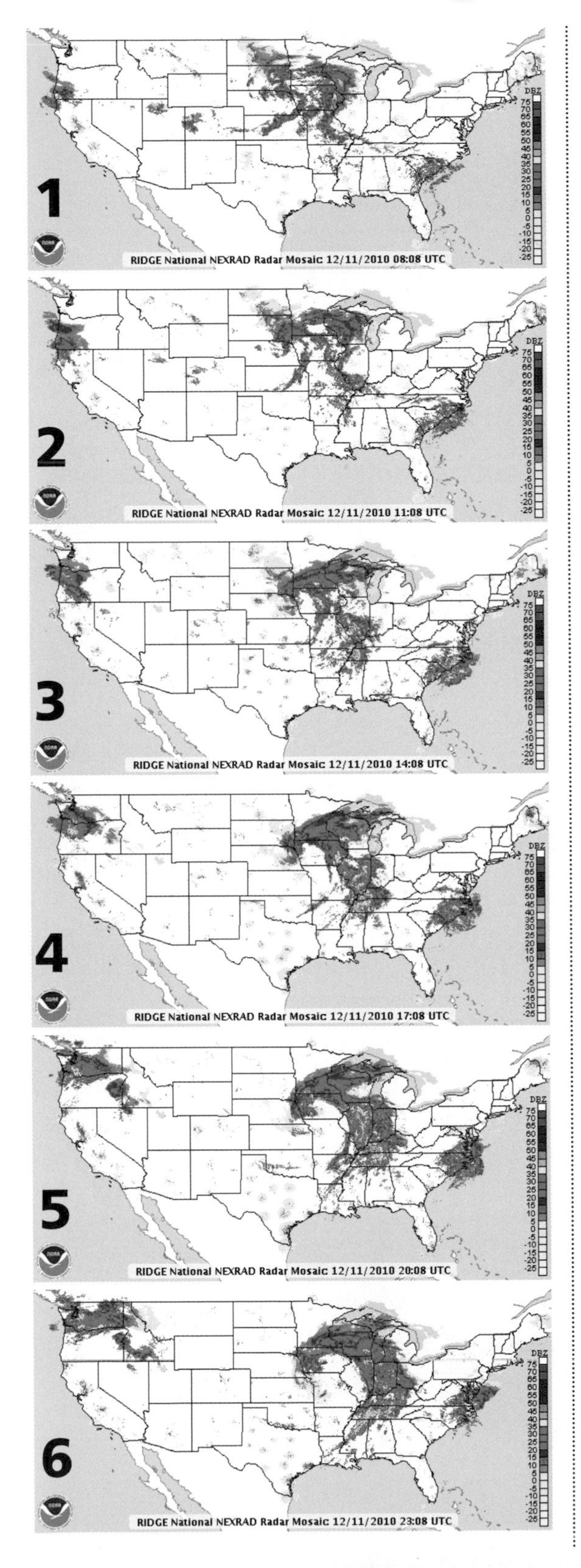

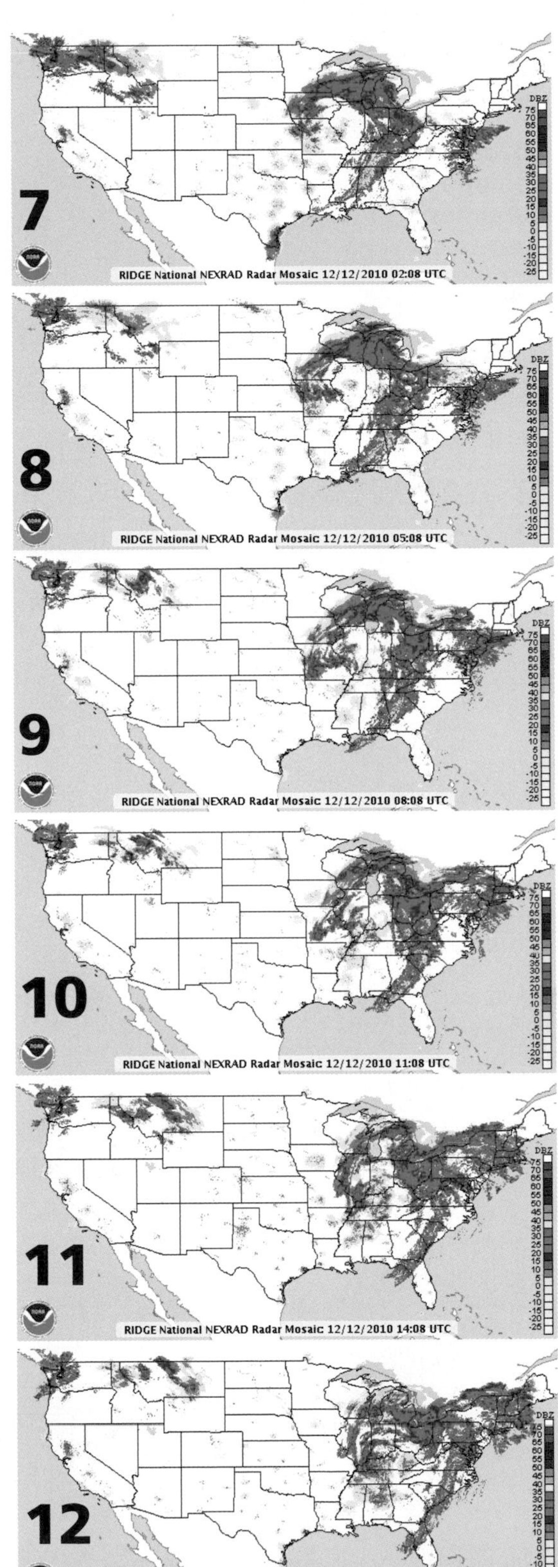

Raindrops and Cloud Droplets

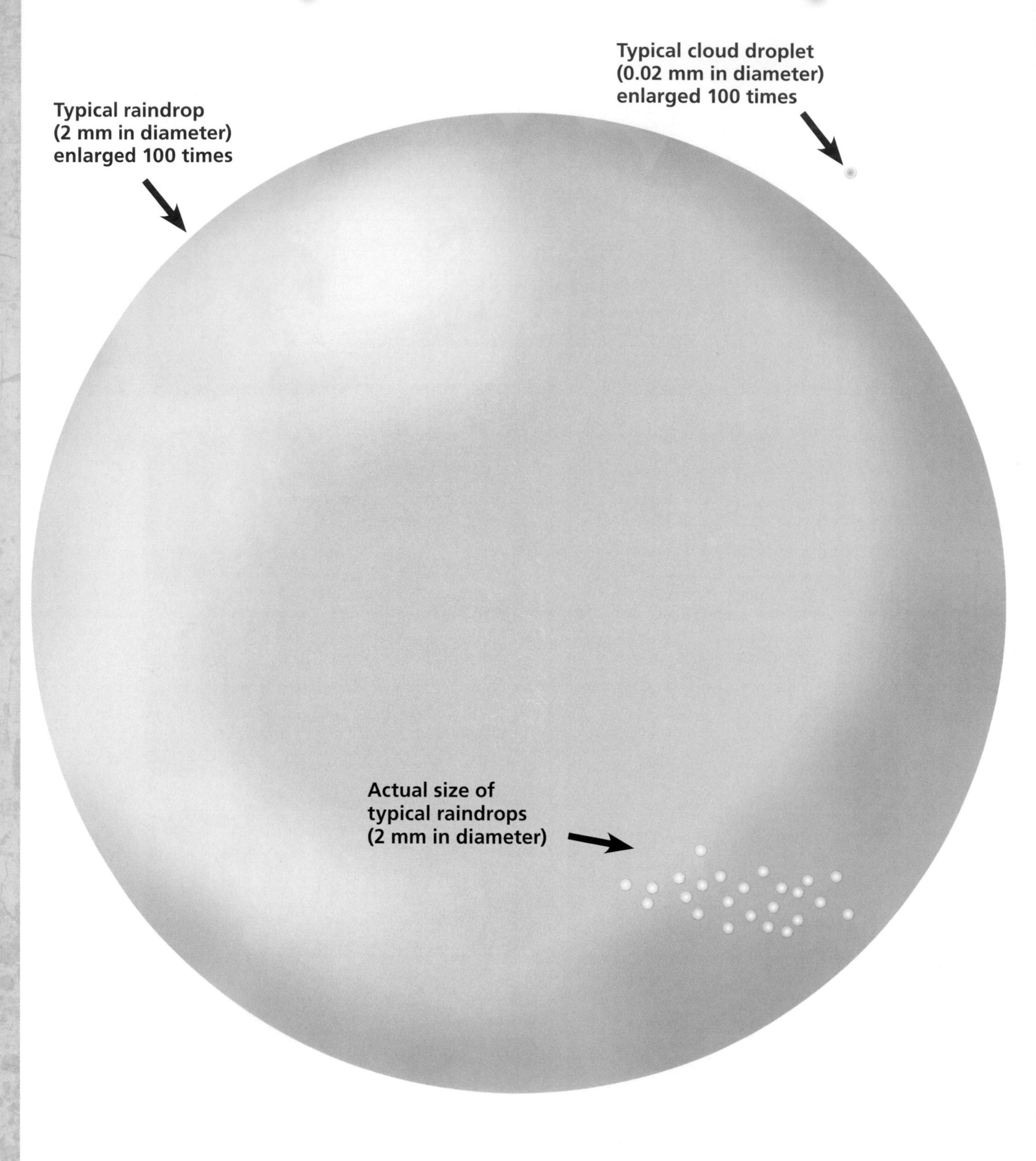

Atmospheric Data Comparison

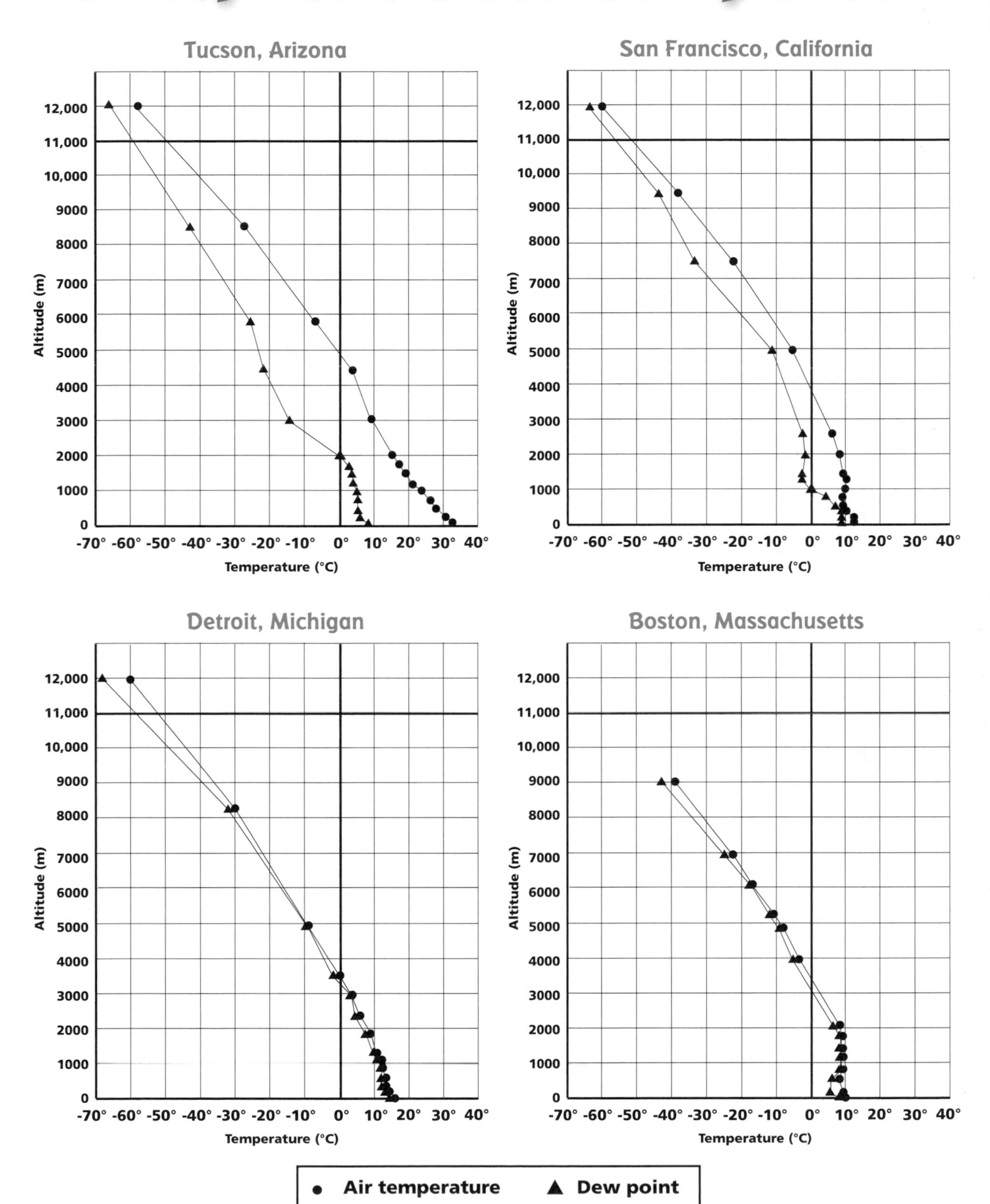

Fronts

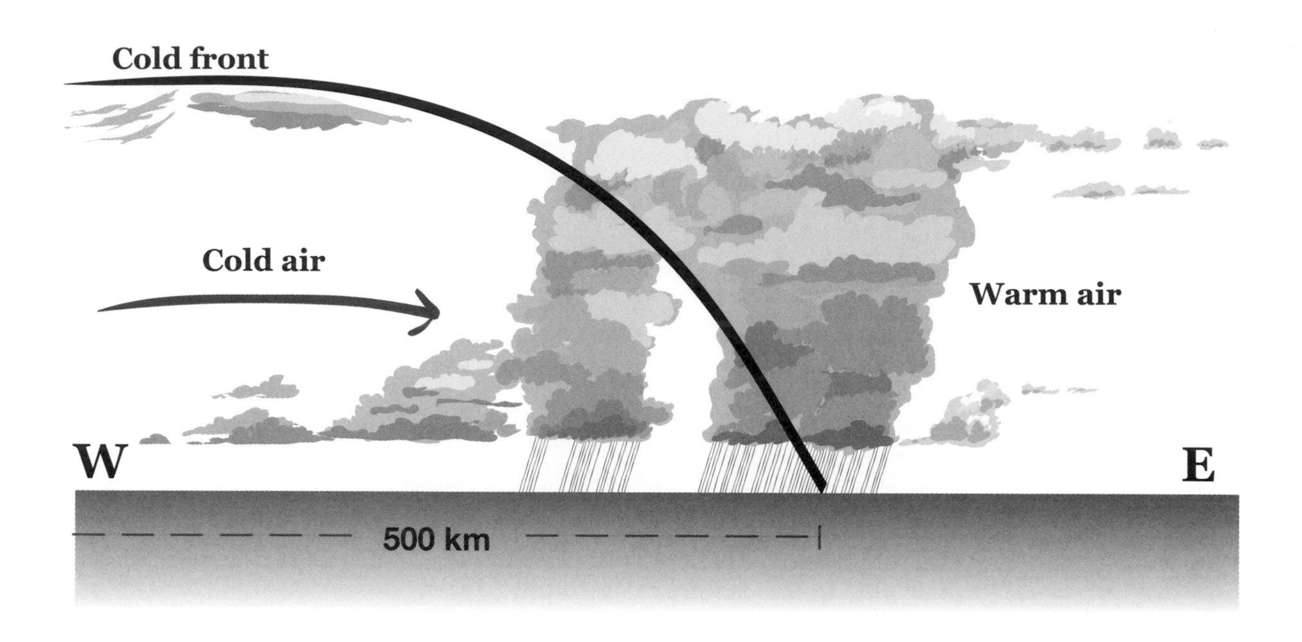
Cold front
Cold air
Warm air
W
E
500 km

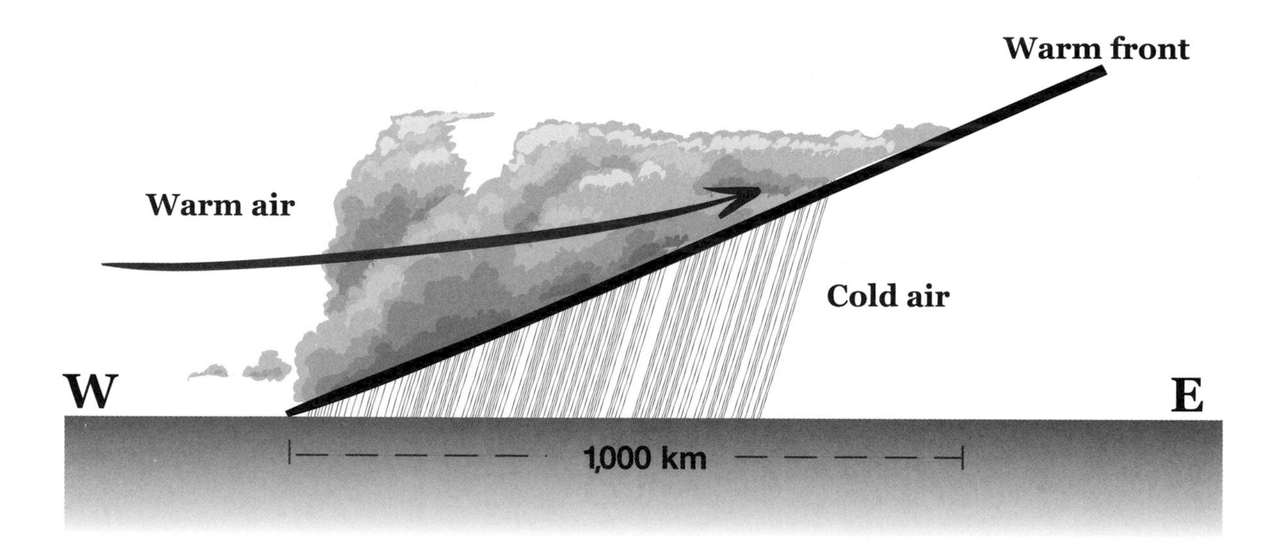
Warm front
Warm air
Cold air
W
E
1,000 km

Weather and Fronts

Cold Front

Weather observation	Before front passes	While front passes	After front passes
Winds	South to southwest	Gusty, shifting	West to northwest
Temperature	Warm	Sudden drop	Drops steadily
Pressure	Falls steadily	Small drop, then sharp rise	Rises steadily
Clouds	Cloud cover increases; cirrus, cirrostratus, cumulonimbus	Cumulonimbus	Cumulus
Precipitation	Short period of showers	Heavy rain, sometimes with thunderstorms, including hail	Showers followed by clearing
Visibility	Fair to poor, hazy	Poor, then improving	Good, unless there are showers
Dew point	High and steady	Drops sharply	Continues to lower

Warm Front

Weather observation	Before front passes	While front passes	After front passes
Winds	South to southeast	Variable	South to southwest
Temperature	Cool/cold, slowly warming	Steadily rising	Warming, then steady
Pressure	Usually falling	Leveling off	Rises slightly, then falls
Clouds	Clouds usually in this order: cirrus, cumulostratus, altostratus, nimbostratus, stratus, then fog; cumulonimbus also possible in summer	Stratus and variations of stratus	Clearing followed by stratocumulus; cumulonimbus also possible in summer
Precipitation	Light to moderate rain or drizzle	Drizzle or none	Usually none, but sometimes light rain or showers
Visibility	Poor	Poor, but improving	Hazy or fair
Dew point	Steady rise	Steady	Rising, then steady

US Map with Western Cities

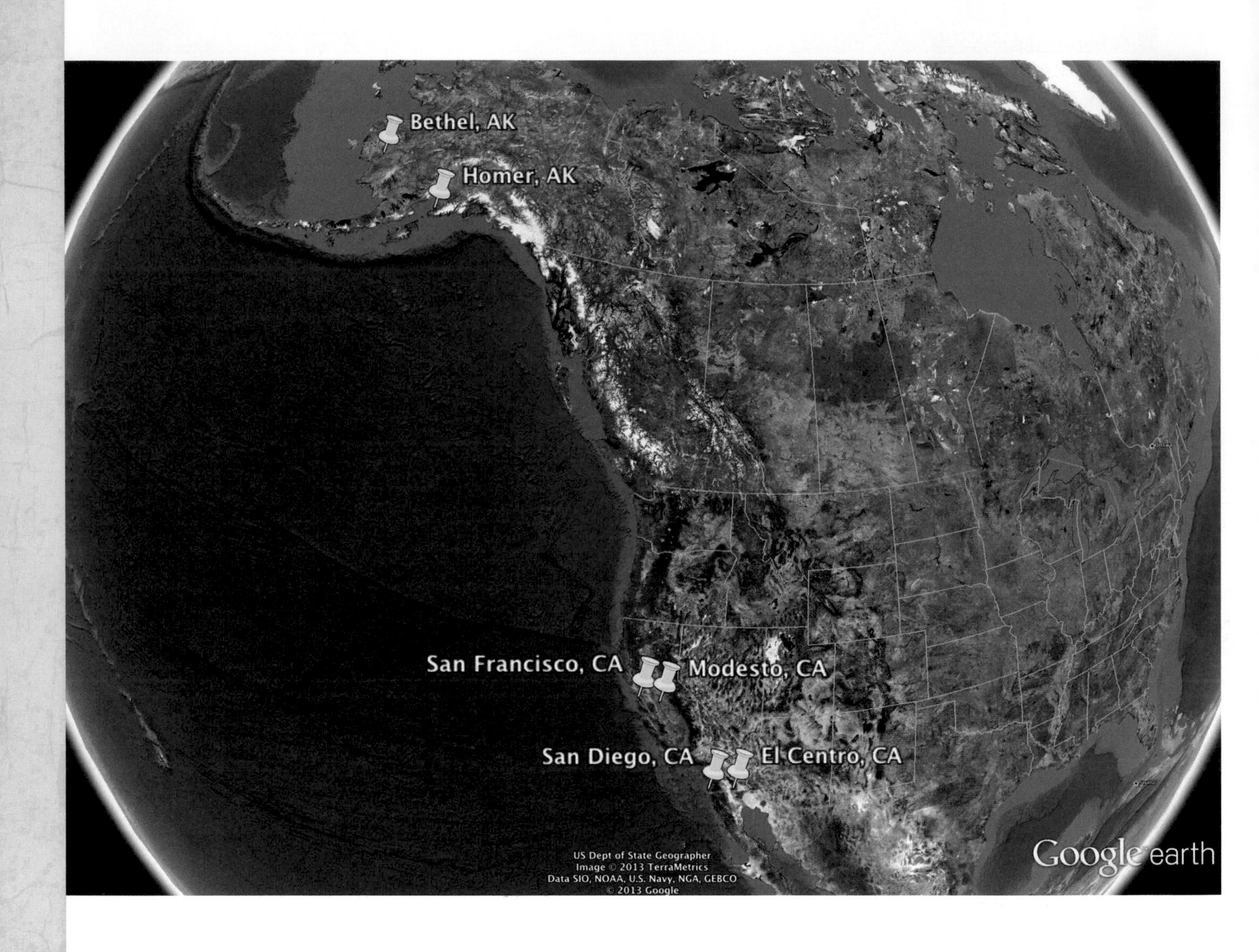

Alaska Climate over 30 Years

Homer, Alaska

Latitude: 59.6° N
Longitude: 151.5° W
Elevation: 10 m
Distance from ocean: 0 km
Average annual precipitation: 64.6 cm

Average temperature in °C

Jan	Feb	Mar	Apr	May	Jun
-5.0	-4.2	-1.7	2.2	6.7	10.0

Jul	Aug	Sep	Oct	Nov	Dec
12.2	12.2	8.9	3.1	-1.4	-3.3

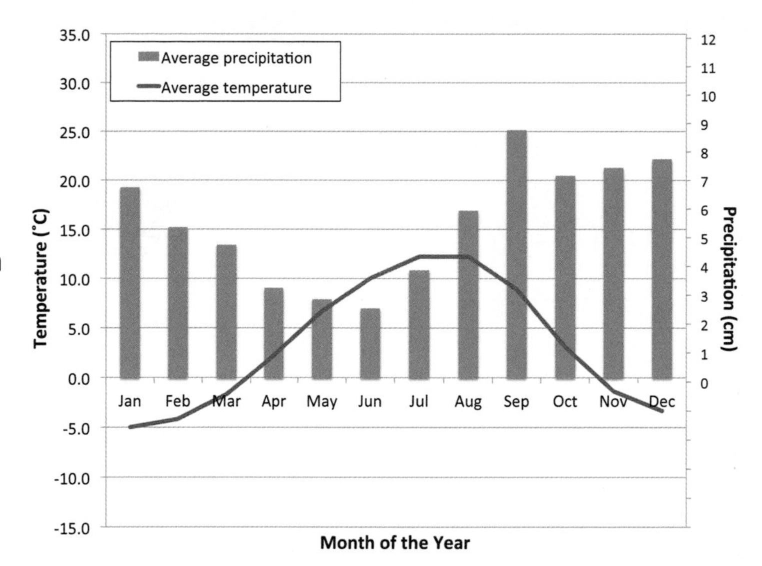

Bethel, Alaska

Latitude: 60.8° N
Longitude: 161.8° W
Elevation: 10 m
Distance from ocean: 96 km
Average annual precipitation: 38.8 cm

Average temperature in °C

Jan	Feb	Mar	Apr	May	Jun
-14.2	-13.6	-9.7	-3.6	5.0	10.6

Jul	Aug	Sep	Oct	Nov	Dec
13.3	11.9	7.5	-1.1	-8.1	-12.5

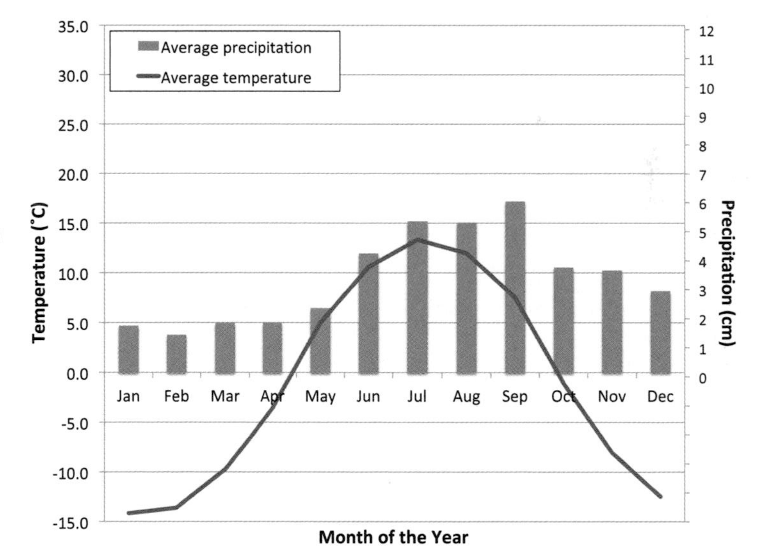

Southern California Climate over 30 Years

San Diego, California

Latitude: 32.7° N
Longitude: 117.2° W
Elevation: 9 m
Distance from ocean: 0 km
Average annual precipitation: 30.4 cm

Average temperature in °C

Jan	Feb	Mar	Apr	May	Jun
13.1	13.3	13.9	15.3	16.9	19.2

Jul	Aug	Sep	Oct	Nov	Dec
21.7	22.2	21.7	18.9	15.6	13.1

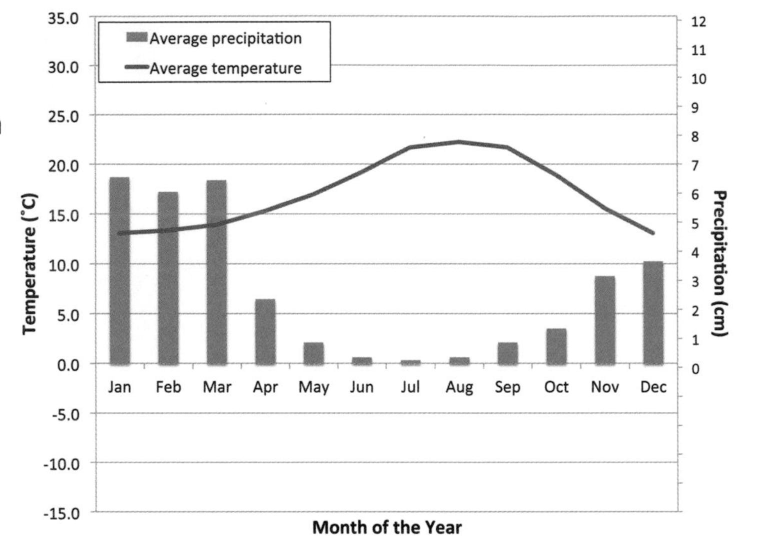

El Centro, California

Latitude: 32.8° N
Longitude: 115.6° W
Elevation: -9 m
Distance from ocean: 144 km
Average annual precipitation: 7.5 cm

Average temperature in °C

Jan	Feb	Mar	Apr	May	Jun
13.1	15.3	17.8	20.8	25.3	29.7

Jul	Aug	Sep	Oct	Nov	Dec
33.1	33.1	30.0	23.9	16.9	12.8

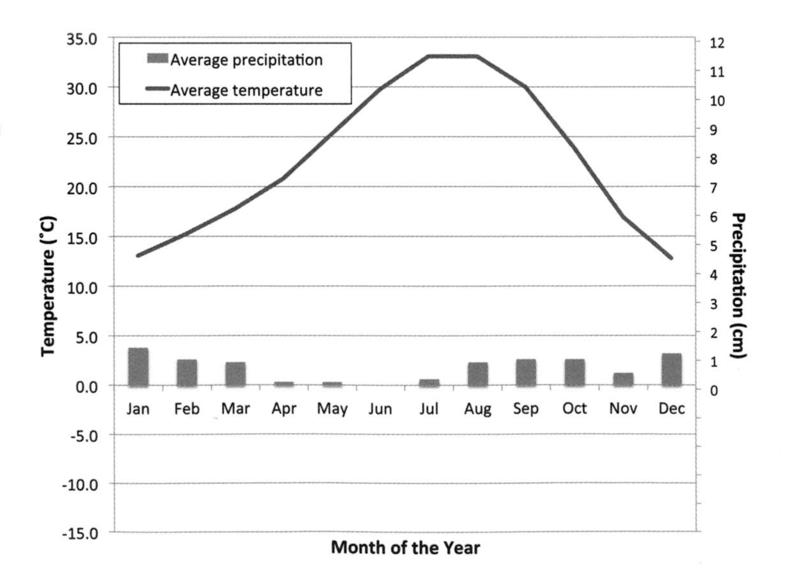

Northern California Climate over 30 Years

San Francisco, California

Latitude: 37.8° N
Longitude: 122.4° W
Elevation: 9 m
Distance from ocean: 0 km
Average annual precipitation: 56.5 cm

Average temperature in °C

Jan	Feb	Mar	Apr	May	Jun
11.1	12.8	13.1	14.2	14.4	15.8

Jul	Aug	Sep	Oct	Nov	Dec
16.1	16.9	17.5	16.9	14.2	11.7

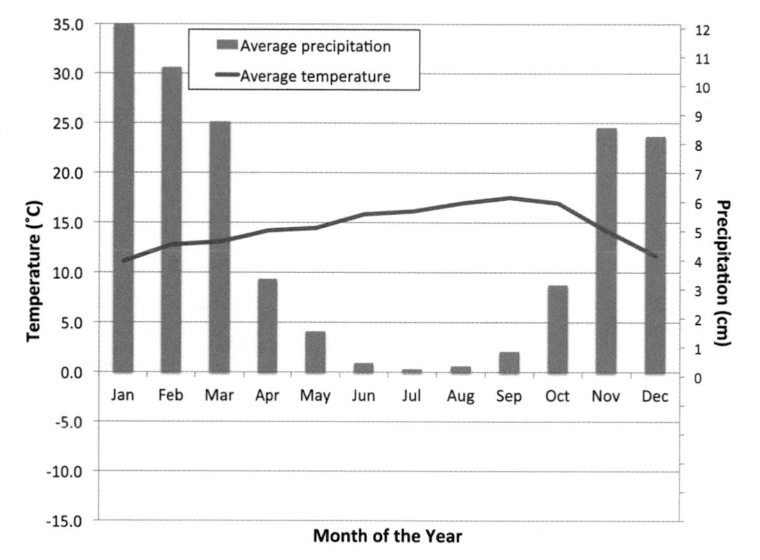

Modesto, California

Latitude: 37.6° N
Longitude: 121.0° W
Elevation: 24 m
Distance from ocean: 129 km
Average annual precipitation: 33.3 cm

Average temperature in °C

Jan	Feb	Mar	Apr	May	Jun
8.3	11.7	13.6	16.4	20.0	23.3

Jul	Aug	Sep	Oct	Nov	Dec
25.6	24.7	22.8	18.3	12.2	8.1

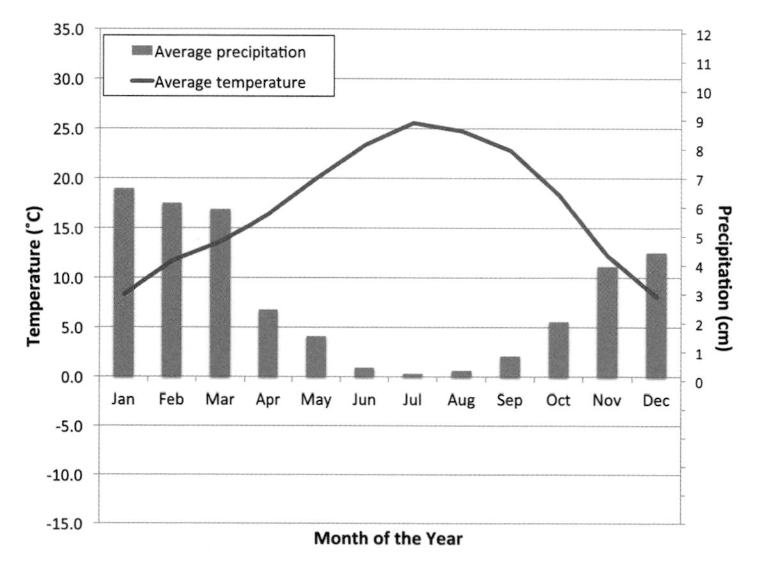

Climate Graph A

Climate Data for Denver, Colorado

1901–1930

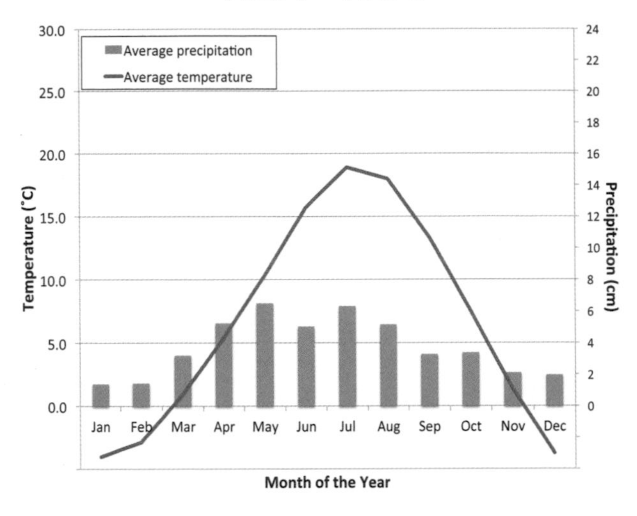

1981–2010

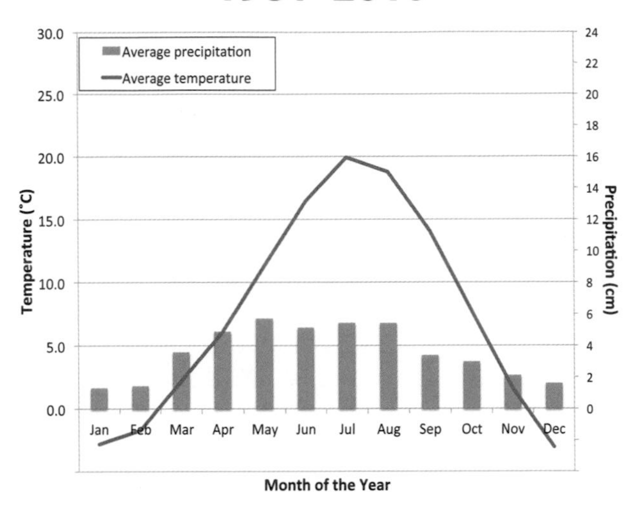

Climate Data for Miami, Florida

1901–1930

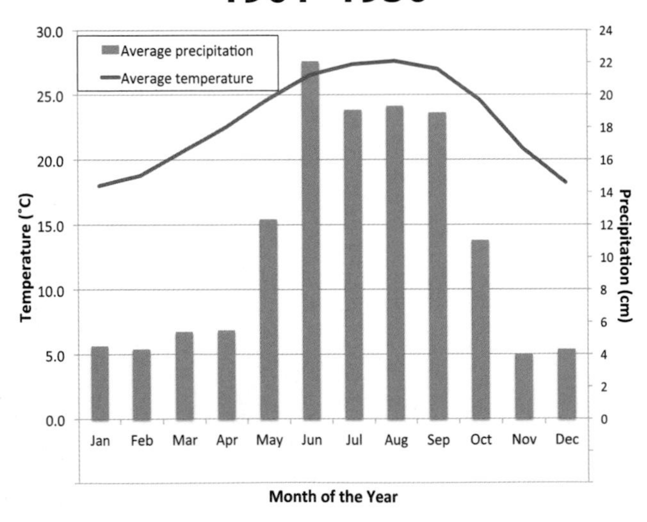

1981–2010

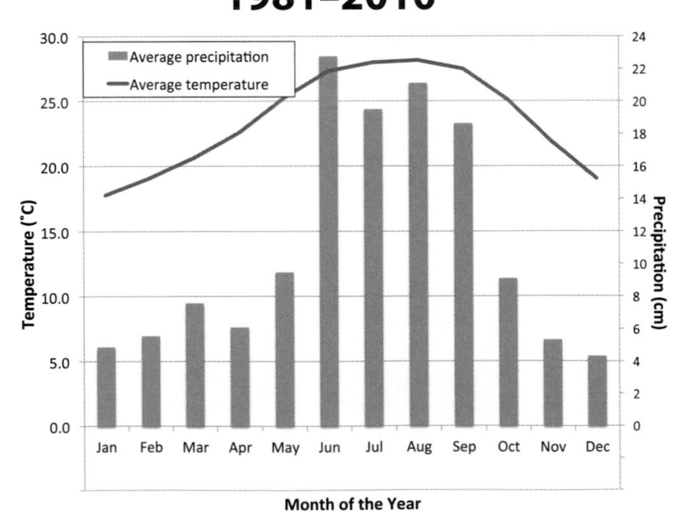

Climate Graph B

Climate Data for Los Angeles, California

1901–1930

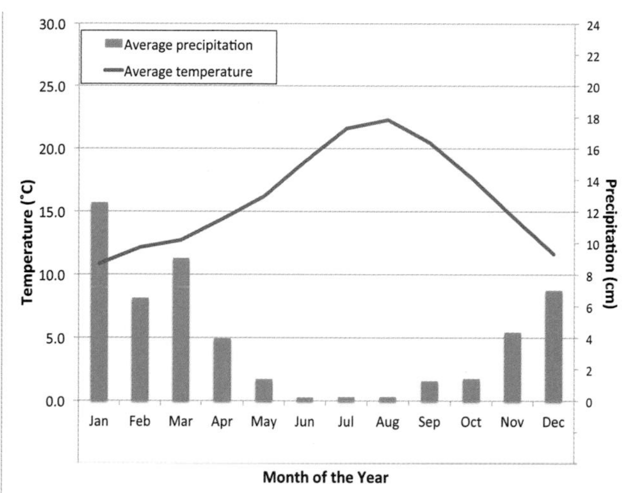

1981–2010

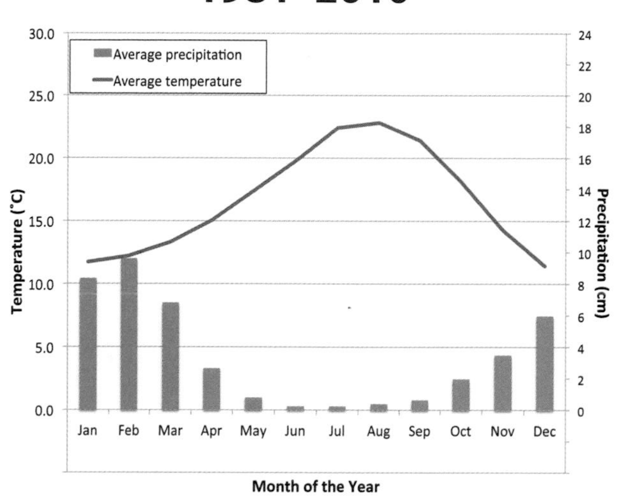

Climate Data for New York, New York

1901–1930

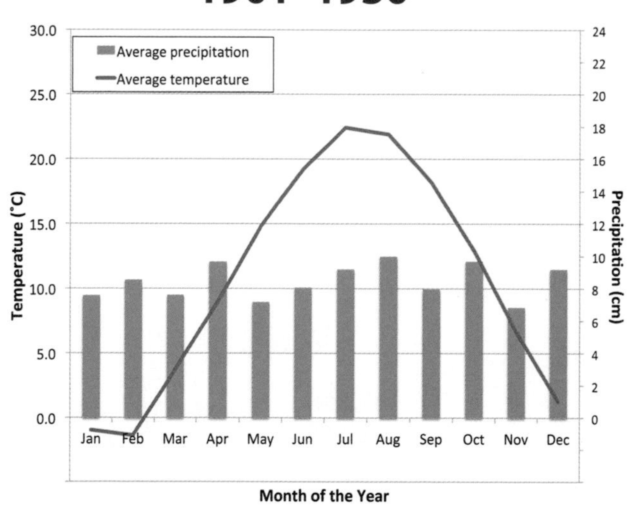

1981–2010

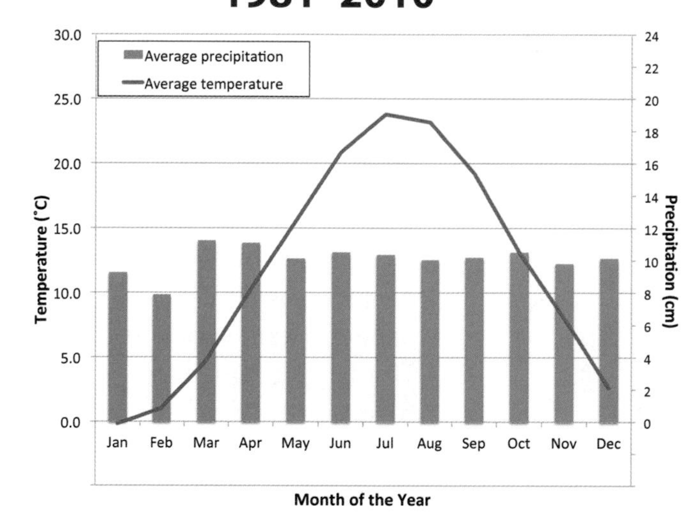

Science Safety Rules

1. Always follow the safety procedures outlined by your teacher. Follow directions, and ask questions if you're unsure of what to do.
2. Never put any material in your mouth. Do not taste any material or chemical unless your teacher specifically tells you to do so.
3. Do not smell any unknown material. If your teacher asks you to smell a material, wave a hand over it to bring the scent toward your nose.
4. Avoid touching your face, mouth, ears, eyes, or nose while working with chemicals, plants, or animals. Tell your teacher if you have any allergies.
5. Always wash your hands with soap and warm water immediately after using chemicals (including common chemicals, such as salt and dyes) and handling natural materials or organisms.
6. Do not mix unknown chemicals just to see what might happen.
7. Always wear safety goggles when working with liquids, chemicals, and sharp or pointed tools. Tell your teacher if you wear contact lenses.
8. Clean up spills immediately. Report all spills, accidents, and injuries to your teacher.
9. Treat animals with respect, caution, and consideration.
10. Never use the mirror of a microscope to reflect direct sunlight. The bright light can cause permanent eye damage.

Glossary

absorb to soak up

air the mixture of gases surrounding Earth

air pressure the force exerted on a surface by the mass of the air above it; also called atmospheric pressure

atmosphere the layer of gases surrounding Earth. Its layers include the troposphere, stratosphere, mesosphere, thermosphere, and exosphere.

atmospheric pressure the force exerted on a surface by the mass of the air above it; also called air pressure

axis an imaginary axle that a planet spins on

barometer a weather tool that measures air pressure

bimetallic strip a narrow band made of two metals stuck together

blizzard a severe storm with low temperatures, strong winds, and large quantities of snow

boundary current a large-scale ocean current along the coastline

carbon dioxide (CO_2) a greenhouse gas found in Earth's atmosphere. Carbon dioxide is created by natural and human-made processes.

cold front a zone where a faster-moving cold air mass collides with a warm air mass, resulting in brief, intense precipitation

compass a tool that uses a free-rotating magnetic needle to show direction

condensation nuclei tiny surfaces on which water may condense

conduction transfer of energy from one place to another by contact

convection heat transfer in a fluid in which hot fluid rises and cold fluid sinks, setting up a cycle

convection cell mass of fluid flowing in a cycle in an area

cumuliform a type of cloud that is puffy, sometimes fast-moving and rapidly growing

density the amount of mass in a material compared to its volume

doldrums the calm area around the equator

downburst a small, very intense downdraft

drought a less-than-normal amount of rain or snow over a long period of time

dust devil a small rotating wind that becomes visible when it collects dust and debris

dust storm a severe weather condition in which strong winds carry dust over a large area

El Niño a flow of unusually warm water in the eastern Pacific Ocean that causes many changes in weather in other places (such as a lot of rain in areas that are usually dry)

equinox a day of the year when the Sun's rays shine straight down on the equator

exosphere the layer of the atmosphere above the thermosphere. The exosphere makes the transition from the atmosphere to space.

flash flood a short, rapid, unexpected flow of water

flood a large amount of water flowing over land that is usually dry

fluid substance that can flow, such as a gas or liquid

global warming a warming trend on Earth that affects evaporation and precipitation

greenhouse gas a gas that absorbs and radiates heat energy in the atmosphere, effectively trapping heat

groundwater water stored below Earth's surface

Gulf Stream a boundary of ocean currents in the North Atlantic Ocean

gyre a large system of rotating ocean currents

Hadley cell a huge convection cell that covers much of Earth at the equator

hail precipitation in the form of small balls or pellets of ice

horse latitude the windless area around 30 degrees north of the equator

hurricane a cyclone or severe tropical storm that produces high winds in the Northern Hemisphere

hygrometer a weather tool that measures humidity

infrared radiant energy not visible to humans that is found beyond the red end of the visible spectrum

jet stream a narrow band of high-altitude wind that affects weather conditions on Earth

kinetic energy energy of motion

land breeze a wind that blows from land to sea

lightning a flash of light caused by a discharge of static electricity between two clouds or from a cloud to Earth

mass the amount of matter in something

matter anything that has mass and takes up space

mesosphere the layer of the atmosphere above the stratosphere

methane (CH_4) a variable gas in the atmosphere that is also a greenhouse gas

microburst a small, very intense downdraft

nitrogen (N_2) a colorless, odorless gas that makes up about 78 percent of Earth's atmosphere

North Star the reference star pointed to by Earth's North Pole

ocean current a global water pattern affected by winds, differences in water density, and tides

orbit the path and length of time one object takes to travel around another object

oxygen (O_2) a gas that makes up about 21 percent of Earth's atmosphere. Oxygen is used by all plants and animals during cellular respiration.

ozone (O_3) a form of oxygen that forms a thin layer in the stratosphere

permanent gas a gas in the atmosphere in which the amount of gas stays constant. Oxygen (O_2) and nitrogen (N_2) are permanent gases.

photosynthesis a process used by plants and algae to make sugar (food) out of light, carbon dioxide (CO_2), and water (H_2O)

planet an object that orbits a star and is massive enough for its own gravity to force it into a spherical shape

prevailing wind predictable wind produced by the combination of high- and low-pressure areas and the Coriolis effect

radiant energy energy that travels through air and space

radiosonde an instrument sent into Earth's atmosphere to measure temperature, air pressure, and humidity

revolution traveling around something

rip current a local current that moves extremely fast

rotation spinning on an axis

salinity the amount of salt concentration

sea breeze a wind that blows from sea to land

season a period of the year identified by changes in hours of daylight and weather

severe weather out-of-the-ordinary and extreme weather conditions

solar energy heat and light from the Sun

solstice a day of the year when Earth's North Pole is angled farthest either toward or away from the Sun

star a large, hot ball of gas

step leader a downward path of electric charge from a cloud to Earth

straight-line wind a strong wind that has no rotation

stratiform a type of cloud that is flat and layered

stratosphere the layer of the atmosphere above the troposphere. The ozone layer is in the stratosphere.

thermometer a weather tool that measures temperature

thermosphere the layer of the atmosphere above the mesosphere

thunder loud, explosive sound created by lightning

thunderstorm severe weather that results from cold air flowing under a warm, humid air mass over the land

tornado a rapidly rotating column of air that extends from a thunderstorm to the ground. Wind speeds can reach 400 kilometers (km) per hour or more in a tornado.

trade wind a wind from between 5 degrees and 25 degrees north or south latitude

troposphere the layer of the atmosphere that begins at Earth's surface and extends upward for an average of 15 kilometers (km). Weather happens in the troposphere.

typhoon a cyclone or severe tropical storm that produces high winds in the Pacific north of the equator and west of the International Date Line

variable gas an atmospheric gas whose amount changes based on the environment. Carbon dioxide (CO_2), methane (CH_4), and ozone (O_3) are variable gases.

warm front a zone where a faster-moving warm air mass collides with a cold air masss, resulting in prolonged, light precipitation

water vapor (H_2O) the gaseous state of water; a gas that is found in Earth's atmosphere

weather factor a property of weather, such as temperature, humidity, air pressure, or wind

wildfire a fire occurring in nature that can be driven by winds

wind meter a weather tool that measures wind speed with a small ball in a tube

Index